INTRODUCTION TO PRACTICAL ORE MICROSCOPY

LONGMAN EARTH SCIENCE SERIES

Edited by Professor J. Zussman and Professor W. S. MacKenzie,
University of Manchester

B. C. M. Butler and J. D. Bell: Interpretation of Geological Maps
P. R. Ineson: Introduction to Practical Ore Microscopy
P. F. Worthington: An Introduction to Petrophysics
B. W. D. Yardley: An Introduction to Metamorphic Petrology

INTRODUCTION TO PRACTICAL ORE MICROSCOPY

P. R. Ineson

215 28

Longman
Scientific &
Technical

Copublished in the United States with
John Wiley & Sons, Inc., New York

Longman Scientific & Technical
Longman Group UK Limited,
Longman House, Burnt Mill, Harlow,
Essex CM20 2JE, England
and Associated Companies throughout the world.

Copublished in the United States with
John Wiley & Sons, Inc., 605 Third Avenue, New York, NY 10158

First published 1989

British Library Cataloguing in Publication Data

Ineson, P. R.
  Introduction to practical ore microscopy.
  – (Longman earth science series)
  1. Ores 2. Microscope and microscopy
  I. Title
  553 TN265

ISBN 0-582-30140-8

Library of Congress Cataloging in Publication Data

Ineson, P. (Peter)
  Introduction to practical ore microscopy.

  (Longman earth science series)
  Bibliography: p.
  Includes index.
  1. Ores – Optical properties. 2. Optical mineralogy.
  3. Microscope and microscopy. I. Title. II. Series.
  QE390.I53 1989   552'.06   88–3015
  ISBN 0-470-21072-9 (USA only)

Set in 10/12 pt Linotron 202 Ehrhardt Roman

Produced by Longman Singapore Publishers (Pte) Ltd.
Printed in Singapore

CONTENTS

PREFACE AND ACKNOWLEDGEMENTS

The study of the ore minerals, by whatever means, relies especially on the experience and help of a teacher. This book could not have been produced in its present form without the help and advice of colleagues around the world for which I am most grateful; the following are particularly thanked: A. M. Evans, P. Garrard, C. Halls, J. McMahon Moore, A. H. Rankin, F. J. Sawkins and S. G. Walters, as well as the instrument manufacturers who kindly provided text illustrations and technical information.

I have also gained immensely over the years from teaching students in the UK and the USA and from the feedback they have provided.

Professor J. Zussman provided continual and invaluable assistance in reading the manuscript and suggesting major and minor modifications. However, any errors and imperfections which remain in the text are the author's responsibility. I would be pleased to receive communications with respect to these in the hope that a future edition can thereby be improved.

P. R. INESON
June, 1988

We are indebted to these copyright holders for permission to reproduce the following tables:

Elsevier Science Publishers B. V. for table 4.5 (Picot & Johan 1982) from pp. 28–34; Geological Survey of Canada for tables 4.6 & 4.9 (McLeod & Chamberlain 1968); Pergamon Press for table 5.2 (Ramdohr 1980) from Vol. 1, pp. 185–93; John Wiley & Sons Inc. for table 7.1 (Craig & Vaughan 1981) from pp. 31–2.

1 INTRODUCTION

Ore microscopy is the traditional name for the study of opaque minerals using a polarising 'incident light' microscope. It is also known as reflected light microscopy. It has applications in the fields of mineralogy, economic geology, mineral dressing, metallurgy and in the study of igneous, metamorphic and sedimentary rocks which contain opaque minerals. Although mineral identification is an important aspect of the study of the opaque ores, an examination of the textures and structures is also valuable.

This book is designed to form an introduction to the subject. The reader who wishes to study the subject in greater depth is referred to the more comprehensive texts by Cameron (1961), Shouten (1962), Freund (1966), Galopin and Henry (1972), Uytenbogaardt and Burke (1973), Ramdohr (1980), Craig and Vaughan (1981) and Picot and Johan (1982). However, the principal methods of identification and the most commonly noted textures are here described for a selection of minerals which should suffice for most students. Company geologists can examine either run-of-the-mill or exotic mineral grades, whereas students at universities, technical colleges, mining schools, museums and similar institutions normally see only exotic or high-grade samples, and although these can be helpful in 'learning the trade' they can be a misleading preparation for operational conditions. This has been borne in mind in making the selection of minerals to be dealt with in the present text.

Ore microscopy and the use of the reflected light microscope with its varied ancillary equipment cannot be considered to be a comprehensive study. It frequently needs to be combined with other investigations, e.g. chemical analyses (by spectrographic, X-ray fluorescence, electron microprobe, atomic absorption, wet chemical and other techniques), X-ray diffraction, universal stage, and grain size measurements. In general, ore microscopic examination is a useful, or even an essential, precursor to such investigations. Likewise, the ore minerals cannot be fully comprehended without a knowledge of the deposits in which they occur. There are many suitable texts which cover this aspect, but the following may be found to be appropriate: Park and MacDiarmid (1964), Brown (1967), Stanton (1972), Baumann (1976), Wolf (1976), Barnes (1979), Evans (1980), Jensen and Bateman (1981) and Hutchison (1983).

Mineralogical texts are also plentiful but the following may prove to be sufficient at this initial level of study: Deer, Howie and Zussman (1962), Ribbe (1974), Rumble (1976) and Vaughan and Craig (1978).

2 THE ORE MICROSCOPE

2.1 GENERAL COMMENTS

The microscope (Fig. 2.1) used for the study of 'ore' or 'opaque' minerals is basically the same as a petrological microscope except that it uses reflected light. (It should not be assumed, however, that experience with the petrological microscope can be directly transferred to the use of the reflected-light microscope.) Some modern microscopes (Fig. 2.2) may be dual purpose and have a combination of both transmitted and reflected light sources built into their assembly. These are capable of examining polished samples, petrological thin sections and polished thin sections. They are particularly useful when both the opaque and non-opaque minerals have to be examined for their mutual relationships and textures, e.g. ore minerals together with gangue or the host/wallrock minerals.

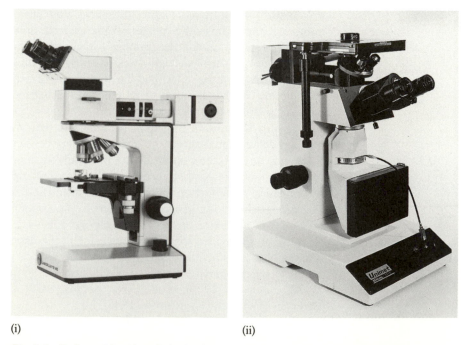

(i)   (ii)

Fig. 2.1 Reflected/incident light (ore) microscopes. (i) Leitz 'Laborlux 12ME'. Reproduced by courtesy of Ernst Leitz Wetzlar GmbH. (ii) Union 'Unimet' inverted stage microscope with binocular head. Reproduced by courtesy of Vickers Instruments Ltd.

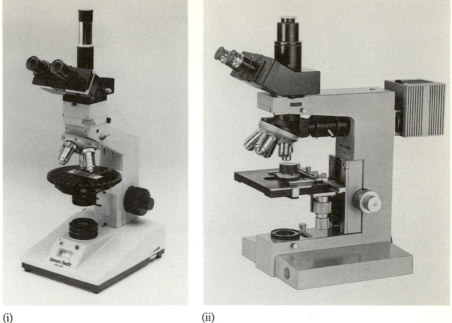

(i)   (ii)

Fig. 2.2 Dual-purpose microscopes capable of viewing samples in transmitted as well as reflected/incident light. (i) 'MP3500' James Swift Polarising Microscope with trinocular head. Reproduced by courtesy of Prior Scientific Instruments Ltd. (ii) Leitz 'Metallux II' microscope. Reproduced by courtesy of Ernst Leitz Wetzlar GmbH.

The microscope should possess:

1. A rotatable stage.
2. One or more objective lenses of either the 'dry' (or air) type or the 'oil' immersion type.
3. An ocular (eyepiece) lens.
4. Polariser and analyser.
5. A reflector.
6. A light source, preferably built-in and stabilised.

It may also have the following attachments or accessories:

1. A sample holder (various models are available).
2. A mechanical stage.
3. An eyepiece micrometer.
4. A photometer.
5. A monochromator and/or filters.
6. A camera attachment.

2.2 THE ROTATABLE STAGE

The stage (Fig. 2.3) must be perpendicular to the light path and centred relative to the objectives. Once correctly aligned, it does not normally need any further

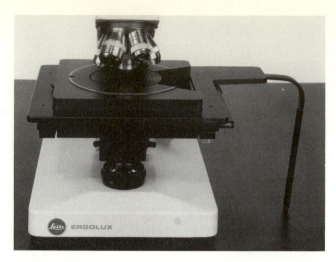

Fig. 2.3   The Leitz 'Ergolux' Microscope stage. Reproduced by courtesy of Ernst Leitz Wetzlar GmbH.

adjustments. Angular measurements are possible with the verniers often provided. A mechanical stage may be fitted for the systematic traversing of a sample in the *X* and *Y* directions.

2.3   THE OBJECTIVE LENS(ES)

These may be classified as: achromatic, apochromatic and fluorite or semi-apochromatic and are applicable for both air and oil immersion. They may also be made to give a flat primary image and as such are prefixed 'flat-field' or 'plan' and, if so, are very useful for examining large fields of view and when photographing the sample. The magnification (degree of enlargement) can vary from ×2 to ×125 or greater. The numerical aperture (NA) indicates the ability of the lens to distinguish fine structural detail and determines the depth of field and the useful range of magnification. A typical lens 'turret' is shown in Fig. 2.4.

Achromatic lenses are the most common and the cheapest. They are corrected for spherical aberration for one colour, normally yellow-green, and for chromatic aberration for two colours. When used with white light, colour fringes may be seen in the outer image margins. If black-and-white microphotography is used with these lenses, fuzziness is possible. In monochromatic light, the image is sharp.

Apochromatic lenses are better but more expensive. They are corrected for spherical aberration for two colours (blue and green) and for chromatic aberration for the primary spectral colours. They produce a sharper image and should give a better print in colour photomicroscopy. To give an even better performance, they should be used with 'compensating' eyepieces.

Fluorite objective lenses are a compromise between achromatic and apochromatic lenses, and should be used with compensating eyepieces.

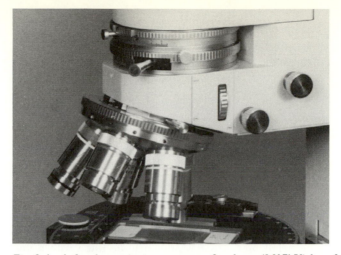

Fig. 2.4 A five-lens microscope turret fitted to a 'M17' Vickers Microscope. Reproduced by courtesy of Vickers Instruments Ltd.

Most lenses (Fig. 2.5) of a low to medium magnification are 'dry' lenses, and as such are designed to be used when air is the medium between the sample and the objective lens and/or when the specimen is covered with a glass. Though they are frequently used in transmitted light studies, in reflected light microscopy they may give distorted images. Immersion objective lenses, which are manufactured to give all possibly required magnifications, should be the most commonly used type in reflected light studies and especially so when high resolution and magnification are demanded. They are, however, not often used by persons with little experience because they require frequent and careful cleaning of the lens

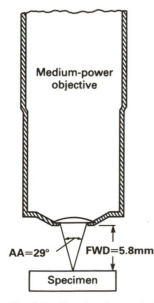

Fig. 2.5 Sketch of a typical lens used in ore microscopy. It shows the free working distance (FWD) and angular aperture (AA). Reproduced from *An Introduction to the Methods of Optical Crystallography* by F.D. Bloss (1961); Holt, Rinehart and Winston Inc.

(to stop it gathering dust) each time either the lens or the sample is changed. The immersion medium may be oil or water, but it is normally oil with a refractive index of 1.515. These special lenses reduce the reflectance of the mineral, increase the colour differences, reduce the diffuse light scattering and permit the observation of weak anisotropism and bireflectance.

2.4 THE OCULAR (EYEPIECE) LENS(ES)

Monocular and binocular eyepieces are available for most microscopes. If possible, to alleviate eye strain, binocular eyepieces should be used and the position of the two eyepieces adjusted so that a single perfectly circular field of view is observed. The oculars are normally of × 5 and × 10 magnification, with or without a perpendicular crosshair, but may be fitted with a micrometer disc or grid for the accurate measurement of parts of the sample. These internally fitted attachments are capable of being brought into focus by rotating the eyepiece. This should always be one of the very first operations to be undertaken. Photomicroscopy eyepieces do not, normally, have crosshairs and are of the compensating type in order to correct for chromatic aberration.

2.4.1 CARE AND CLEANING OF LENSES

The objective and ocular lenses are prone to gathering dust as well as accidentally touching the sample or your eyes. They may thus become 'dirty' or produce an effect called 'blooming'. The lenses should not be cleaned with ordinary tissues or cloths, or breathed upon, since this apparently rapid and easy remedy may cause one of the most expensive pieces of the microscope, i.e. the lens, to be ruined. The recommended lens-cleaning tissues should be used or assistance requested from an experienced person.

2.5 THE POLARISER AND ANALYSER

The polariser is normally located between the light source and the objective lens and often between two sets of diaphragms. It is usually a polaroid plate which, is capable of being easily rotated. It is mounted perpendicular to the analyser. The polarisation effects are often more easily noted if the polars are a *few degrees* from the true 90° position, and this is especially so for anisotropic minerals **if** they occur in a groundmass or matrix of a more strongly anisotropic mineral. This effect may be achieved by *slightly* (3–5°) rotating either the polariser or the analyser from the totally crossed position.

Rather than rotating the stage, for the movement of the whole sample may distract the eyes, it is better or more advisable to rotate the analyser or polariser about the extinction position. The presence or absence of anisotropism in the grain may thus be confirmed. If need be, adjust one of the diaphragms, in order to view only that grain.

2.6   THE REFLECTOR

The reflector, which is either a glass plate or a prism, is the means by which the light is brought vertically onto the polished samples surface. A glass plate reflector only allows about 25% of the illuminating light that first reached the reflector to ultimately reach the ocular lens. A prism reflector has approximately 50% efficiency. If the sample is viewed in a conoscopic field of view, only half of the polarisation figure will be seen, for the other half is occupied by the prism. Normally these reflectors are fixed and adjustment is a specialised operation.

2.7   THE LIGHT SOURCE (ILLUMINATION, ILLUMINATOR)

Possibly one of the most frequent errors in ore microscopy is not to pay sufficient attention to the alignment of the light source.

This aspect cannot be overstressed, for the light source should provide optimum conditions of adequate light, evenly distributed throughout the field of view and be comfortable to the eyes. Variations in the colour of the same mineral in the same sample when viewed with different microscopes are largely due to the incorrect alignment of the light source, rather than slight differences which may occur in the effective colour temperature of the light source. The actual colour temperature is not a vital aspect in routine investigations but is vital in photomicroscopy due, in the main part, to the various requirements of the different photographic films which may be used. It is also important in any reflectance measurements as well as determinations of the correct colour of a mineral, as the colour is a function of the light source.

Of the various and important aspects which should be carefully and systematically checked when the illumination is set up, the diaphragms are essential. The illuminator aperture diaphragm is used to restrict the field of view as well as reducing stray scattered light. The illuminator field diaphragm, which controls the angle of the cone of light on the specimen, should be adjusted to just enclose the field of view. It eliminates elliptical polarisation and minimises the contrast. If a third diaphragm is fitted to the microscope, it may be adjusted to sharpen the image.

The light source may consist of a bulb which, if it is old, may be operating at too low a wattage or even be partially misaligned – an all too common feature – even if the source is in-built and correct alignment is assumed to be the case. It must be adjusted in accordance with the microscope manufacturer's instructions. A number of factors may give rise to one of the most frequent errors during the initial alignment procedures: that is the appearance of filament images and/or coloured filament-shaped zones in the field of view. Radical adjustments need to be made and it may require a service before the microscope can be used. For routine work, a tungsten filament lamp operated by a rheostat is adequate. A frosted glass screen may help to eliminate the filament image but should not be inserted to remove this effect, for the effect itself must be corrected. Likewise, these bulbs often have a bias towards the yellows and reds, and some people therefore insert a pale blue filter between the lamp and the rest of the illumination system. This actually changes the colour balance but not the temperature. It is frequently used in transmitted-light microscopy for it provides a more daylight-like colour temperature. However, it should *only* be used after all the alignment procedures noted have been undertaken. If it is possible, use a high-intensity halogen filament incandescent lamp or a xenon discharge lamp in order to provide a better light, see Fig. 4.1.

2.8   SAMPLE HOLDER

The sample holder ensures that the polished surface is perpendicular to the incident light beam. A simple press that embeds the sample onto a piece of plastic or plasticine mounted on a glass slide or some such flat plate will suffice. The sample may, however, be machined so that it has flat and parallel upper and lower surfaces, or more sophisticated sample holders may be used. They are called the Lanham and Leppington Stages. The Leppington Stage (Fig. 2.6) is mounted on the microscope's stage. It consists of a circular five-position sample holder which, by holding the samples against an upper flat surface, ensures that all the samples are automatically in focus when each is rotated into the field of view. Hence, it is highly suitable for reflectivity measurements and cross-referencing when rapid movement from one sample to another is required. All the holders must achieve the same requirements.

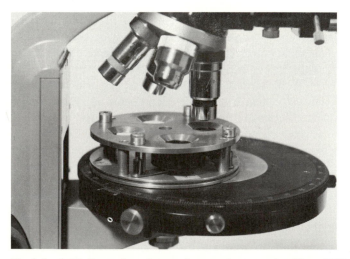

Fig. 2.6   A McCrone Leppington four-station, auto-levelling, 360° rotating stage. Reproduced by the courtesy of McCrone Research Associates Ltd.

2.9   MECHANICAL STAGE

Mechanical stages which are attached to the rotatable stage hold the sample securely, and advance it in the *X* or *Y* direction in variable incremental amounts.

They are invaluable in the systematic traversing of the sample when such operations as point or grain counting are required. Cumulative mineral percentages and modal analyses can thus be obtained.

2.10   EYEPIECE AND STAGE MICROMETERS

Eyepiece micrometers inserted into one of the binocular eyepiece tubes or the monocular lens tube are capable of rotating and traversing one free crosshair relative to a fixed crosshair. The movement of the free graticule is measured on an attached vernier dial. When combined with a stage micrometer, they are capable of measuring the absolute size of any feature on the sample. They are standard pieces of equipment in hardness testing where they determine the size of the indented pits.

A stage micrometer is subdivided into a millimetre or micrometre (μm) scale and enables the estimation of grain size as well as the calibration of the ocular scales or grids. The micrometers are small mounted and inscribed metal discs, positioned and viewed in the same way as the polished samples. They also find applications in studies of textures, mill products and other industrial applications where grain sizes have to be determined.

2.11   PHOTOMETER

A photometer (see Fig. 4.1) may be in-built or attached to a standard microscope by means of a special head adapter which, normally, mounts the photometer vertically above the microscope and allows visual examination of the field by means of a secondary inclined ocular tube. It is used to measure the reflectance of a mineral grain. To obtain accurate results, it must be used with a stabilised and high-intensity light source, monochromators and reflectance standards.

2.12   MONOCHROMATORS AND FILTERS

The optical properties of minerals can vary with wavelength, and it may therefore be necessary to provide light of a specific or variable wavelength. Monochromators are designed to fulfil this requirement. There are two types of monochromator: the fixed interference filter and the continuous-spectrum type. The former passes light within a specific band width (less than 15 nm if 'narrow', and 15 to 50 nm if of the 'broad' band type). A specific and separate filter is required for each wavelength. The continuous-spectrum monochromator is an interference filter the wavelength of which varies along its length. A window, the width of which may be varied, may be moved along the filter to give the wavelength required. It provides monochromatic light over the visible spectrum and beyond.

A monochromator may be built-in or external. Whichever type, it is located in the light path, usually after the light source and before the photometer.

3 SAMPLE PREPARATION

3.1 GENERAL COMMENTS

A sample suitable for examination should be a plane polished surface as free as possible of pits, scratches, fractures, pores, thermal and mechanical modifications and relief. This requirement is often achieved, by manual or mechanical means, on a range of materials. However, on ore samples the aim is rarely attained. Various methods (Otto and Rensburg 1968), all with minor modifications, are available, but essentially they all necessitate that the sample is carefully selected, trimmed, cast, ground and polished. Routine examination is normally performed on *polished sections*; however, polished thin sections and doubly polished thin sections, structurally etched sections, structurally etched thin sections and grain mounts can also be prepared.

3.2 PREPARATION

Normally a sample is cut with a water-lubricated, thin-bladed diamond saw. The lubricating fluid prevents, or aims to prevent, heating of the specimen while it is being cut. If the specimen is fractured, porous or friable or even consists of grains or powder, it will be necessary to impregnate it initially. Vacuum impregnation of epoxy resin or plastic into the sample is the standard technique.

Most samples are cast in circular or rectangular moulds which may have an enclosing permanent plastic ring (Fig. 3.1). The ring gives near uniformity to the overall thickness and permits ease of storage, labelling and handling.

The casting process should retain the mineralogy and textural relationships. The older techniques, as described by Cameron (1961: 36–42), used Bakelite and resulted in the sample being subjected to heat and pressure. These techniques are still used for metallurgical samples but as they modify the low-temperature or hydrous minerals and may cause twinning or exsolution (even at 100 °C), the use of cold-setting epoxy resins or plastics is recommended for the preparation of all samples.

The cut sample which frequently has one flat surface lying on the bottom of the mould or jig (Fig. 3.2) is cast in such a manner as to exclude air bubbles. The resultant block should require only the minimal amount of grinding in order to expose a suitable surface for polishing.

(i)

(ii)

Fig. 3.1 (i) A selection of epoxy resin mounted polished samples (with and without enclosing rings) and polished thin sections. (ii) The McCrone Ore Mineral Set. Reproduced by courtesy of McCrone Research Associates Ltd.

3.3 GRINDING AND POLISHING

Grinding removes surface irregularities and adhering resin, reduces the overall thickness and removes any surface deformations which may have resulted from cutting, and it prepares the smooth surface for subsequent polishing. It is essential, therefore, that grinding does not itself cause any surface deformation (by

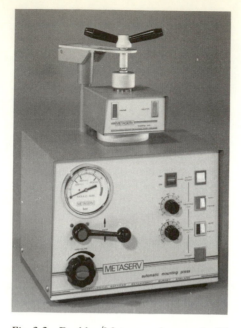

Fig. 3.2 Buehler/Metaserv Automatic Mounting Press. Reproduced by courtesy of Metallurgical Services Laboratories Ltd.

pressure or heating), introduce contaminants (from the grinding medium) and result in plucking of the mineral surface which may contaminate the lap and cause the consequential scratching of subsequent samples. The compounds used in grinding are designated in *grit*, *mesh* or *micron* sizes (Table 3.1). Fixed abrasives (emery, diamond) embedded in metal, epoxy or paper are often preferred to loose abrasives for they plane off a uniform surface.

Table 3.1 Comparative table of grit, mesh and micron sizes of abrasives used in grinding and polishing

Grit or mesh size	Microns	Inches
100	150	0.0059
200	75	0.0030
260	60	0.0024
325	45	0.0018
400	38	0.0015
600	30	0.0012

Polishing is a continuation of grinding but with the progressive use of finer and finer abrasives. The final removal of the surface layer of deformation and a resultant scratch-free surface are the aims.

The rotating laps of the polishing equipment (Fig. 3.3) may be smooth or grooved. Grooved laps require resurfacing, but are capable of giving relief-free surfaces. The laps themselves may be constructed from cast-iron, aluminium, copper or lead and different laps may be used at different stages in polishing.

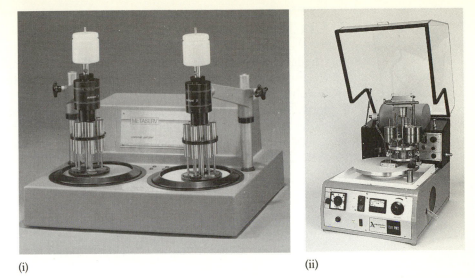

(i)   (ii)

Fig. 3.3   (i) Buehler/Metaserv twin-wheel polishing machine with automatic attachments. Reproduced by courtesy of Metallurgical Services Laboratories Ltd.   (ii) Logitech's 'WG2 Polishing System'. Reproduced by courtesy of Logitech Ltd.

3.3.1   VIBROPOLISHING

Vibropolishing, another variant, is an efficient and effective final-stage process. It is only successful when applied to an initially uniform flat sample surface. It does, unfortunately, produce a high relief if the sample contains relatively hard and soft mineral components.

3.3.2   ELECTROLYTE POLISHING

Electrolyte polishing removes material from the sample electrically and mechanically. The electrolytic fluid, commonly also an etchant, must be chemically reactive, so it may remove some mineral phases, e.g. silver or bismuth.

3.4   POLISHED THIN SECTIONS

As most ores contain both opaque and transparent minerals, a thorough investigation necessitates the use of transmitted and reflected light microscopy, as well as the electron microprobe. This requirement has resulted in techniques being developed for the preparation of thin sections (~0.03 mm thick) that are polished on the upper uncovered surface, or on both surfaces (doubly polished thin sections). By this method, certain textures are detectable which cannot be seen in normal polished sections (i.e. cast and polished samples), in 'petrological' thin sections and even in polished (single-sided) thin sections. Typical examples are the textures exhibited by sphalerite, cassiterite, rutile, cinnabar and tetrahedrite-tennantite.

3.5 ETCHING

Chemical etching has been used for mineral identification (Short 1940). The electron microprobe has largely replaced this method but it still has its uses in revealing structure. Ramdohr (1980) indicates that it is suitable for detecting zoning, twinning and grain boundaries and for enhancing mineral phases when modal analyses are required. For further information on this method, see Chapter 8 – Mineral Tables – and the specific results which are noted with respect to the individual minerals.

3.6 GRAIN MOUNTS

Mineral grains may be studied by casting them in an epoxy resin or plastic mould and cutting and polishing the sample as described previously.

3.7 SAMPLE STORAGE

Considering all the time, care, attention and cost involved in producing a sample suitable for examination, it is folly to ruin it by incorrect storage. For research purposes, each sample may need to be prepared, examined and photographed immediately it has been produced. Normally, however, most people store some or all the samples for a considerable time and undertake repolishing when they

Fig. 3.4 Buehler/Metaserv Desiccant Specimen Storage Cabinet. Reproduced by courtesy of Metallurgical Services Laboratories Ltd.

have tarnished, oxidised or undergone some secondary time-related alteration. In order to minimise these secondary processes, the samples are frequently stored in a dust-free and dry environment. Suitable storage cabinets are commercially available which partially exclude dust and air. They can be equipped with a desiccation medium (silica gel) and have rubber-sealed doors (Fig. 3.4). They also enable the polished samples to be collated, labelled and handled without damage. The storage cabinet shown in Fig. 3.4 is capable of holding 300 or more samples.

The upper polished surface of a specimen should never be wiped clean with a cloth prior to examination. Only, and then very infrequently, an alcohol-impregnated lens tissue may be used to remove oil or dust adhering to the surface. Finger prints resulting from careless handling are a common cause of coating the surface with 'oil'. As certain of the ores tarnish naturally with time, this property may be used as a means of identification. For example, silver invariably tarnishes and may require buffing prior to examination. The use of proprietary silver cleaning agents may be considered as a quick and easy polishing method.

The sample should *never* be stored with the polished surface downwards, even on tissue or cloth, since these may have become covered with dust or abrasive material and repolishing will be needed.

4 MINERAL IDENTIFICATION

4.1 SIMPLE OPTICAL TECHNIQUES

4.1.1 INTRODUCTION

It is assumed that only a standard reflecting microscope is available without modifications or additional equipment. If so, this can be an advantage, for it can be tempting to try to use the more complex techniques prior to comprehending and mastering the more basic methods. In the case of ore microscopy, as compared with the study of thin sections in transmitted light, the properties are qualitative (for the most part) and there is no substitute for experience. Time spent on examining as many different minerals as possible from as many different localities and environments will lead to proficiency with the technique.

As well as a polished specimen, it may help to have at hand a sample of the material which has not been cut, mounted and polished. The Mineral Tables (Ch. 8) give the general colour, streak and lustre for each mineral in hand specimen. Reference to any text on descriptive mineralogy will provide far more information with respect to the megascopic features. This section of the Mineral Tables may be useful for a coarse-grained sample or intergrowth, but is virtually useless for very fine-grained samples, because the colour and other properties of the specific mineral may be masked by the host rock or adjacent minerals.

For the observation of a mineral's optical properties, the microscope can be used in two principal modes:

1. Without the analyser, i.e. using only the polariser (which is generally left permanently inserted). This is said to be viewing the sample using linear or 'plane polarised light', and is the mode used for observations of:

   Colour
   Reflectance
   Bireflectance and Reflectance pleochroism

2. With the analyser inserted (at 90° to the polariser). This is said to be examining the sample using 'crossed polars' and is the mode used for observation of:

   Isotropism v. Anisotropism
   Polarisation colours (Rotation properties)
   Internal reflections

Observations may be made using air and/or oil immersion lenses.

4.1.2 COLOUR

The colour of a mineral, when viewed under plane polarised light or between crossed polars, is one of the most important properties; unfortunately, it is one of the most difficult to use when starting to learn ore microscopy.

The following check list may help to overcome some of the common problems:

1. That the microscope is correctly set up.
2. That the sample is freshly polished or, if not, has been kept in as dry an atmosphere as possible. If it has not, then there may be difficulties, because certain minerals (e.g. native copper and silver) tarnish readily when left in air and need frequent repolishing in order to show their 'correct' colours.
3. That the samples are being viewed under suitable and constant illumination. The microscope should be set (if possible) with as high an illumination as is comfortable to the eyes.
4. That you are not colour blind. A large number of people are partially colour blind and never realise it. A friend can be asked to look at the sample and to describe what colours are seen; accept, however, that the description of a colour may vary somewhat from person to person.

As the majority of minerals have colours which vary from pure white to dull grey and only a small number are strongly coloured (Table 4.1), it is necessary to differentiate subtle colour differences within the pastel shades as well as varying shades of grey and white to off-white. The following advice may help:

1. View as many samples of the *same* mineral as possible and hence see the variations which one mineral can show.

2. Make a note of the colour observed and do not rely completely on the tables in this or any other book, since they can only be a guide. In some circumstances, the sample being examined may not show any of the colours that the table lists.

3. While viewing one mineral, choose one which covers the whole field of view and alter the illumination. Variations in the intensity of the colour should be seen, e.g. from bright to pale yellow or even from bright yellow to white. Reset the illumination to what it was originally. This will illustrate the problems which can occur if a constant intensity of illumination is not maintained even when observing different samples of the same mineral.

4. Appreciate that the variation in colour depends on a mineral's surroundings, i.e. it may be set in a non-opaque country rock or next to minerals which are brightly coloured or duller in colour. This effect is called 'neutral colour interference'. These differences can vary depending on whether the sample is 'in air' or 'in oil'. The Mineral Tables (Ch. 8) are designed to help with this phenomenon. As an example see page 102 on chalcopyrite, in the section headed 'colour'. The '→' indicates 'chalcopyrite next to . . .', so chalcopyrite next to galena may appear to be more yellow than if it is on its own, or next to gold/silver when it may appear to show a colour which is duller or greenish yellow.

Likewise, if a mineral occupies the whole field of view, it may appear to give a totally different colour than when it occurs as minute grains associated

Table 4.1   Colours exhibited by the ore minerals

Predominant colour (Relative)		Isotropic (or weakly anisotropic)	Anisotropic
1. STRONGLY COLOURED	Blue	Chalcocite	Covellite
	Yellow	Gold	Chalcopyrite
			Millerite
			Cubanite
	Red/brown	Bornite, copper	
	Pink/purple/violet	Bornite, copper	
2. SLIGHTLY COLOURED (in comparison with adjacent minerals)	Blue	Tetrahedrite	Psilomelane
			Hematite
			Proustite
			Pyrargyrite
	Green	Tetrahedrite	Stannite
	Yellow	Pyrite, pentlandite	Marcasite
			Niccolite
3. FAINTLY COLOURED (in comparison with adjacent minerals of not too intense a colour)	Blue		Cuprite
			Cinnabar
	Pink/purple/violet	Cobaltite	Niccolite
	Yellow	Pyrite	Marcasite
	Red/brown	Magnetite	Pyrrhotite
			Enargite
			Ilmenite

with other minerals or fragments of the country rock. It is advisable therefore to start with a low-power objective lens in the microscope, thereby viewing as much of the sample as possible, and to have all diaphragms open as far as possible.

Colour is a function of: (a) the index of refraction of the medium between the lens and the sample (normally air); and (b) the transparency of the mineral. Most ores are opaque, whereas most gangue minerals are transparent to translucent. These appear darker when polished as they do not reflect light as well as opaque minerals.

4.1.3   BIREFLECTANCE AND REFLECTION PLEOCHROISM

When a mineral is viewed in plane polarised light and is rotated, a change in the reflectance (brightness) and/or the colour of the mineral may occur.

A change in the reflectance is called BIREFLECTANCE

A change in the colour (or its tint) is called REFLECTANCE PLEOCHROISM

It is important to note that cubic minerals do not normally show either of the above effects whatever the orientation of the grains/crystals. Basal sections of hexagonal and tetragonal minerals also do not normally show bireflectance or reflectance pleochroism.

The bireflectance effect is influenced by:

1. The refractive index (RI) of the medium between the lens and the specimen: the higher the RI of the medium, the higher the bireflectance noted.
2. The orientation of the mineral. A mineral may show anything from zero to maximum bireflectance depending upon its orientation. Some authors attempt to quantify this effect with the use of such terms as 'very weak', 'weak', 'moderate', 'strong' and 'very strong'.

The effect may be seen in certain transparent minerals (often the gangue minerals) which possess a strong bireflectance, i.e. Ca, Mg, Fe and Pb carbonates. It is *not* shown by the main rock-forming minerals.

4.1.3(a) Aids to identification

1. As bright an illumination as is comfortable should be used.
2. A lower-power objective lens should be used if possible.
3. All diaphragms should be opened as wide as possible.
4. The light source should be centred.
5. If it is known that the mineral has moderate to weak bireflectance, an area should be sought in which two or more grains are adjacent to each other. In such a situation, the eye should detect the differences shown by the adjacent grains. If the area being examined is one large grain, then a twin plane will give a similar effect.

The reflectance pleochroism shown by covellite is the most striking since it changes from bluish grey to deep blue. Other minerals which show this property are:

Niccolite: pinkish brown to blue-white
Cubanite: pinkish brown to yellow
Graphite: brownish grey to greyish black

The following, although they display the effect, are more difficult (initially) to recognise:

Molybdenite: white-grey to white
Bismuthinite: white-grey to yellow-white
Pyrrhotite: pinkish brown to brownish yellow
Millerite: shades of yellow

4.1.4 CROSSED POLARS ANISOTROPY

When viewed between crossed polars, anisotropic minerals can show changes in the intensity of reflection (brightness) as well as changes in colour. These colour changes are the polarisation colours. They have not been used, nor reported as frequently as they could have been, but they may be useful as additional aids in identification. However, as with all the properties which involve colour, there are problems of interpretation and recording, among which are:

1. The subjective nature of colour observations and descriptions.
2. The variation of colour with different microscopes.

3.　The variation of colour with the intensity of illumination; a good and constant illumination is essential.

For the observation of the true polarisation colours (Table 4.2) the microscope must be set up with the polars exactly crossed. As this is vital the following procedure should be followed:

(a)　Select a × 20 or × 40 objective lens.
(b)　Focus on a recently polished cubic mineral.
(c)　Remove the eyepiece.
(d)　Insert the analyser.
(e)　A perfectly centred black cross on a white background should be seen. If this is not observed, then the polarisers are not crossed.
(f)　Rotate the analyser back and forth until you observe the black cross.
(g)　Insert the eyepiece and do not adjust either of the polarisers, although it is sometimes useful to deliberately uncross the polarisers slightly (Table 4.2).

Table 4.2　Polarisation colours of the ore minerals

Perfectly crossed polars	*Slightly uncrossed polars*
Grey	
Ilmenite	
Manganite	
Brown-grey	
Pyrargyrite	
Sylvanite	Lighter in colour
Green-grey	
Hematite	
Brown (often said to be yellow-brown)	
Wolframite	
Orange-brown	
Chalcocite	
Pink	
Molybdenite	Pink-white
Orange	
Covellite	
Yellow	
Millerite	Yellow to blue
Graphite	Yellow to orange-yellow
Pyrolusite	Brownish yellow
Green	
Bismuthinite	Yellow-green
Marcasite	
Cinnabar	
Cuprite	Green to blue
Blue	
Cubanite	
Goethite	

Table 4.2 (continued)

Perfectly crossed polars	Slightly uncrossed polars
Purple	
Stannite	Bluish
Argentite	Blue-grey to purple

4.1.4 Observations with crossed polars

If a cubic mineral is viewed under crossed polars, whatever the orientation of that sample, it will remain dark or black on a 360° rotation of the stage. It need not be jet-black, but it *will not* show variations in the intensity and/or colour of illumination. Basal sections of hexagonal and tetragonal crystals also show the same effect. Such minerals or sections are isotropic (Table 4.3).

Polished surfaces of minerals which show a change in the intensity of the colour and/or illumination when they are rotated through 360° are said to be anisotropic, see Tables 4.3 and 4.4. If the sample is anisotropic, it should show four positions (90°, 180°, 270° and 360°) where it is at its darkest, i.e. said to be 'in extinction', and four positions of maximum brightness (45°, 135°, 225° and 315°).

Anisotropy can vary in intensity, and the terms 'very weak', 'weak', 'moderate', 'strong' and 'very strong' are frequently used. Note that not only the opaque minerals can show anisotropism but certain carbonates (calcite, dolomite, siderite and cerussite etc.) as well as sphene show anisotropism in reflected light.

Table 4.3 Colours of the isotropic and weakly anisotropic minerals

Colour	Mineral
Red	Native copper
Yellow	Native gold
Cream	Pentlandite
White	Galena
Yellow-white	Pyrite, native silver
Cream-white	Native platinum
Pink-white	Cobaltite
Grey-white	Tetrahedrite, tennantite
Grey	Uraninite, chromite
Light grey	Sphalerite
Brown-grey	Magnetite

4.1.4 Aids to identification

1. Examine as many grains of one mineral as possible. In this way one grain, somewhere on the sample, should be located which shows the maximum degree of anisotropism.
2. For fine-grained ores, high-power objectives and brighter than normal illumination are necessary.

The aperture diaphragm should be nearly closed and the field diaphragm adjusted to cut out all but the field of view being examined.

3.   Rotate the analyser, rather than the stage. Rock it back and forth by some 10° or 20° either side of total extinction.
4.   Compare the effect observed with the notes in the Mineral Tables on pp. 89–148 (Ch. 8).

4.1.4   *Problems*
Anomalous effects

1.   Some cubic minerals are not truly isotropic.
     They may appear weakly anisotropic although this should not hinder their recognition by other features. Pyrite and bornite are especially susceptible to anomalous effects.
2.   If a soft mineral has not been polished carefully, a mesh of fine scratches (often parallel in appearance) remain and may appear to show anisotropism. Carefully focusing under high magnification should enable the distinction between scratches and true anisotropism to be noticed.
3.   During polishing, anomalous anisotropism may be destroyed. If the sample has been prepared with due caution, i.e. knowing the minerals likely to be present and being warned to take extra care, the anomalous anisotropism should be seen in pyrite and other cubic sulphides and arsenides.

4.1.5   INTERNAL REFLECTIONS

Internal reflections are due to light passing through the surface of the mineral and being reflected back from natural or artificial cracks or flaws within it. Samples which have been collected in a mine or open pit exhibit more abundant reflections as they have probably been 'shot-blasted'. Internal reflections can only therefore be detected in minerals which have a degree of transparency and fortunately only a few opaque minerals show this feature. Most of the non-opaque minerals, i.e. the carbonates, sulphates etc., may show abundant internal reflections. Minerals which show this characteristic are noted in the Mineral Tables as well as in a separate listing in Table 4.4. Do not expect to find that every mineral or even every grain of a mineral said to show internal reflections will exhibit the feature. It can sometimes be very difficult to find.
Note: Internal reflections may exhibit a play of colours or one single colour may be observed.

4.1.5   Aids to observation

1.   A low-power objective lens is preferable unless the specimen is very fine-grained.
2.   If possible, the illumination should be set to the maximum but not for too long or it may damage your eyes and reduce the life of the bulb.
3.   For a composite sample or one with small grains, when it may be advisable to use a high-power objective lens, internal reflections are best seen in oil immersion, and it will be necessary to use the special lenses supplied for this technique.
4.   The internal reflections may be best located at the edges of grains.

Table 4.4 Typical internal reflection colours exhibited by some of the ore minerals

Predominant internal colour	Mineral
Yellow	Sphalerite, cassiterite, goethite, rutile
Brown	Sphalerite, cassiterite, rutile, chromite (often red-brown and more distinct in oil), uraninite (rarely seen), wolframite (deep brown; more distinct in oil)
Red	Cuprite, proustite, pyrargyrite, goethite, sphalerite, tennantite, manganite, boulangerite, stibnite (rarely seen), rutile, hematite (more distinct in oil), sphalerite (rare), cinnabar (blood-red)
Colourless	Sphalerite, cassiterite, scheelite, rutile and most of the gangue minerals

Note: Some minerals are listed under more than one colour since they can, and often do, show variations in the colour of their internal reflections.

4.2   REFLECTANCE

4.2.1   GENERAL COMMENTS

The reflectance of a mineral may be quantitatively or qualitatively measured; in other words, measured accurately or estimated by eye. The amount of light which is reflected back to the observer can be influenced by a number of factors.

1. The method used for polishing the sample.
2. Whether or not the sample is coloured. This may be overcome by inserting a filter between the light source and the sample.
3. Whether the sample is tarnished. This normally is due to incorrect storage but it may occur however carefully the sample is stored, i.e. the sample may break down naturally and quickly after polishing
4. The orientation of the sample (see bireflectance). It is best to check numerous grains, if present, and note the variations.

Points 1 and 3 should be overcome by correct polishing and examination after minimum storage.

4.2.2   QUALITATIVE METHODS

As the technique depends on the eye estimating the relative brightness, it is most useful to have a set of reference standards. A small number of minerals showing contrasting reflectances are kept at hand and a ready check made between these and the unknown mineral. In this way the unknown may be said to have a reflectance similar to one of the standards or between two of the standards.

Craig and Vaughan (1981) suggest that the following are used as quick and easy standards:

Quartz and mounting medium	$R\%$	5
Magnetite	$R\%$	20
Galena	$R\%$	43
Pyrite	$R\%$	55

Small differences in the reflectance exhibited 'in air' (the most commonly used and certainly the most easily used medium) are enhanced if the medium used is oil.

4.2.3   QUANTITATIVE METHODS

Until recently the accurate measurement of reflectance was only possible in white light (Galopin and Henry 1972). Now with the development of sensitive photometers and monochromators, we are able to measure reflectance at a specific wavelength. The reflectance (R or $R\%$), the percentage of light reflected back to an observer or instrument, depends upon:

(a) The refractive index of the mineral
(b) The absorption coefficient of the mineral
(c) The wavelength of the incident light

Reflectance may also be directly related to the crystallographic orientation of the sample relative to the plane of polarisation of the incident light.

Whereas cubic minerals have a single value of reflectance ($R\%$) at a specific wavelength of light, uniaxial minerals (hexagonal and tetragonal) have two values of reflectance (R_o and R_e) at a specific wavelength of light, and biaxial minerals (orthorhombic, monoclinic and triclinic) have three principal values of reflectance at a specific wavelength of light, but normally only the minimum and maximum are reported (R_1 and R_2).

As reflectance measurement is the most important quantitative technique used for the identification of an opaque mineral, the reader is recommended to refer to Cameron (1961) and Craig and Vaughan (1981) for more detailed accounts, and to the references cited by them for further study.

4.2.3(a)   The apparatus

1. A light source with stabilised intensity
This is essential and it must be capable of operating at a high filament temperature (3500°C). It is normally capable of having both current and voltage adjusted. Some of the more modern light sources incorporate a sensor which feeds back intensity data so that the lamp is wholly self-correcting.

2. A monochromator
The use of this is optional. It is inserted either between the light source and the sample, i.e. on the end of the polariser/condenser tube or between the two and/or in front of the photocell/photometer (often called the Photometer Head).

3. The diaphragms or stops
There are normally two sets of diaphragms or stops. The first is located on the

polarising/condenser unit and may be of the carousel type (i.e. circular, rotatable and with stops of various diameters). The second set is located just before the photometer and referred to as the measuring field diaphragm (MFD). They limit the beam and the size of the area to be illuminated or measured by the photometer. They also prevent the photometer from being 'blinded', i.e. overloaded.

4. The photometer

This is located on a specially designed eyepiece tube and is more conveniently located vertically above the microscope, for it may need to be rotated in order to be correctly aligned. The viewing eyepiece is capable of being inserted, for viewing the sample, and withdrawn when reflectance measurements are being made.

5. The standards

Normally one standard is used for calibrating the equipment; for accurate work, however, three standards will give a more precise result. The commonly employed standards are:

Standard	Reflectance values (R% – air)			
	R_{470}	R_{546}	R_{589}	R_{650}
Neutral glass (NGl)	4.4	4.3	4.2	4.2
Silicon carbide (SiC)	21.9	20.9	20.2	20.0
Tungsten carbide (WC)	45.5	44.5	43.9	43.6

These have been chosen since they polish well and maintain a good polish. If only one wavelength is to be used, then 546 nm is recommended. If reflectances are to be measured in oil, Cargill D/A oil is recommended; see Galopin and Henry (1972).

The microscope and lamp should be carefully aligned. Mounting all the pieces of the apparatus onto a rigid board maintains the alignment and allows the apparatus to be moved around without the continual need for readjustments.

4.2.3(b) Method of operation

The details will vary for different designs of equipment (Fig. 4.1) but the following is a general guide.

1. The samples under investigation and the standards available should be polished or repolished. If need be, fine scratches can be removed by hand buffing with gamma alumina in water.

2. During the test, the surface of the samples and standards should occasionally be wiped with ethanol on a soft paper tissue.

3. The samples are mounted on a press using a glass slide and some plasticine. It is an advantage if the press has a stop, for all the samples and standards will then be of the same overall thickness. This simple technique is quicker than the use of Lanham or Leppington stages.

4. The samples must be level. This check is vital and is carried out as follows:
   The aperture diaphragm should be closed as far as possible and the

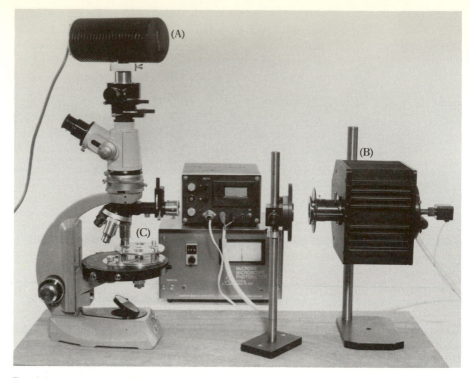

Fig. 4.1 Equipment for reflectance measurements in monochromatic light using McCrone's microscope photometer (A) and intensity-stabilised lamp (B) mounted on a Vickers 'M74c' microscope with a Leppington rotating stage (C). Reproduced by the courtesy of McCrone Research Associates Ltd. and Vickers Instruments Ltd.

Bertrand lens inserted. On rotation of the stage, the image should remain in the centre of the field of view; if not, the image will precess across the field of view. Adjustments can be made by pressing very gently on the side of the sample, or removing it and remounting it using new plasticine.

5. The wavelength of the light required is selected by using the monochromator. This is normally 546 or 589 nm but it can be any of the four COM recommended wavelengths. In fact, any wavelength can be used, but interpretation of the results is then more difficult.

6. An objective lens (×8, ×16, ×20 or ×40 magnification) is selected depending on the granularity of the sample, and the microscope is focused.

7. The field stops, i.e. the photometer and lamp stops, are adjusted. The photometer stop is often adjusted to half the diameter of the lamp stop, but the instructions given with the equipment should be followed.

8. It is advisable to check that the photometer can cover the range of reflectances of the samples and standards.

9. Readings can then be taken with the standard and these should be near to the correct values. If so, five or more readings should then be taken on the sample. The standard should be re-checked and if these are unchanged or nearly so, then the readings for the sample can be accepted; if not, the

sequence should be repeated until reproducibility with the standard is achieved.

10. Tables 4.5 and 4.6 or any of the other published charts and/or tables may be used in conjunction with the measured reflectances to identify the sample. The best sets of data are given on the Cards of the COM (Commission on Ore Microscopy) – see Fig. 4.2 – but the Bowie and Simpson (1980) charts are very good and easy to use.

For cubic minerals, the mean of the two reflectance readings is characteristic of the specimen. For anisotropic minerals, grains which show maximum and minimum reflectance in the wavelength of the light used are chosen. The mean reflectance is measured in each position and these give R_o and R_e or R_1 and R_2.

Table 4.5 Mineral reflectances (%) for monochromatic light

| Mineral | Reflectance | | | | | | | |
| | 420 nm | | 540 nm | | 600 nm | | 700 nm | |
	max.	min.	max.	min.	max.	min.	max.	min.
Silver	71	—	81	—	83	—	86	—
Platinum	58	—	63	—	66	—	68	—
Gold	26	—	63	—	74	—	84	—
Sylvanite	55	41	61	51	62	51	63	55
Millerite	32	35	55	53	58	56	61	58
Marcasite	47	42	55	49	53	49	51	46
Pyrite	40	—	54	—	55	—	57	—
Löllingite	56	50	54	53	54	51	53	49
Arsenopyrite	53	49	52	50	51	51	51	50
Cobaltite	49	—	50	—	53	—	47	—
Bismuthinite	50	40	50	38	48	37	45	36
Pentlandite	35	—	49	—	52	—	56	—
Copper	38	—	48	—	71	—	85	—
Stibnite	54	34	48	33	44	31	40	31
Molybdenite	55	22	45	21	45	20	44	20
Galena	50	—	42	—	41	—	42	—
Cubanite	28	24	41	36	42	39	45	43
Boulangerite	42	40	40	38	40	37	37	35
Pyrrhotite	30	28	37	35	40	38	42	39
Chalcopyrite	18	16	36	35	40	39	42	41
Pyrargyrite	41	35	34	28	31	26	28	23
Tetrahedrite	31	—	33	—	32	—	30	—
Chalcocite	37	—	32	—	30	—	27	—
Proustite	39	37	32	30	30	27	28	25
Hematite	34	31	31	27	29	25	26	23
Pyrolusite	30	19	31	19	30	18	28	18
Cinnabar	36	34	29	28	28	27	27	26
Stannite	22	20	29	27	29	29	27	26
Tennantite	31	—	28	—	26	—	23	—
Cuprite	32	—	26	—	24	—	23	—
Enargite	29	27	26	25	27	26	30	29
Rutile	26	23	24	21	23	20	22	19
Manganite	25	18	22	17	21	16	20	16

Table 4.5 (cont'd)

Mineral	420 nm		540 nm		600 nm		700 nm	
	max.	min.	max.	min.	max.	min.	max.	min.
Covellite	30	16	22	7	19	3	24	22
Bornite	19	—	21	—	26	—	31	—
Magnetite	22	—	20	—	21	—	21	—
Ilmenite	21	19	20	18	20	18	20	19
Graphite	19	8	20	7	22	7	25	8
Wolframite	19	16	17	15	17	15	17	15
Sphalerite	18	—	16	—	16	—	16	—
Uraninite	17	—	16	—	16	—	16	—
Goethite	18	17	14	13	14	12	13	12
Chromite	13	—	12	—	12	—	12	—
Cassiterite	13	12	12	11	12	11	12	11
Scheelite	17	—	11	—	11	—	12	—

This table has been compiled from Picot and Johan (1982) (Reproduced by permission of the publishers: Elsevier and B.R.G.M.)

Table 4.6 Reflectance values for the ore minerals (means and ranges)

Mineral	Reflectance (%)	
	Mean	Range
Argentite	34.2	29.0–36.0
Arsenopyrite	54.7	51.7–55.7
Bismuthinite	45.9	42.0–50.0
Bornite	22.7	21.0–23.9
Boulangerite	40.2	37.0–44.1
Cassiterite	12.0	11.0–12.9
Chalcocite	29.0	18.0–33.5
Chalcopyrite	43.2	42.5–44.0
Chromite	12.7	12.0–14.0
Cinnabar	25.7	24.0–29.0
Cobaltite	53.2	52.0–54.7
Copper	75.8	71.2–81.2
Covellite	14.2	7.0–24.3
Cubanite	40.5	39.2–42.5
Cuprite	27.7	27.1–28.5
Enargite	26.4	24.7–28.1
Galena	43.0	42.4–43.2
Goethite	16.7	16.1–18.5
Gold	73.5	64.0–74.0
Graphite	14.0	6.0–17.0
Hematite	27.3	24.0–30.6
Ilmenite	19.2	17.0–21.1
Löllingite	53.8	53.0–54.7
Magnetite	20.7	20.0–21.1

Table 4.6 (cont'd)

Mineral	Reflectance (%)	
	Mean	Range
Manganite	17.5	14.0–22.0
Marcasite	51.8	48.9–55.5
Millerite	55.4	51.8–60.0
Molybdenite	32.4	15.0–44.9
Niccolite	55.3	52.0–58.3
Pentlandite	51.0	49.6–52.0
Platinum	70.0	————
Proustite	27.3	25.0–28.4
Psilomelane	23.9	23.0–24.4
Pyrargyrite	30.2	28.4–30.8
Pyrite	54.5	54.1–54.8
Pyrolusite	34.4	30.0–41.5
Pyrrhotite	39.3	34.0–45.6
Rutile	22.4	20.0–24.6
Scheelite	11.0	10.0–12.1
Silver	95.1	93.8–96.8
Sphalerite	17.3	16.1–18.8
Stannite	27.6	27.1–28.0
Stibnite	38.0	30.2–47.0
Sylvanite	55.1	48.0–60.0
Tennantite	29.1	28.8–29.7
Tetrahedrite	30.3	29.3–31.2
Uraninite	15.8	14.5–16.8
Wolframite	17.1	15.0–20.2

The data have been reproduced from McLeod and Chamberlain (1968) with the permission of the publishers: Geological Survey of Canada, Department of Energy, Mines and Resources.

4.2.3(c)   **Possible errors**

Galopin and Henry (1972) discussed in detail the possible errors. Briefly there are three main problem areas:

1. *The microscope and the reflectance apparatus.* The better the microscope and in particular the objective lens, the better the results that are likely to be obtained.

2. *Levelling.* This aspect cannot be overstressed. If the microscope is not fitted with a Bertrand lens and the test noted previously cannot be carried out, an equally good alternative is to remove the eyepiece and view the sample in conoscopic light.

3. *Focus.* It is always easier to focus with a lower- rather than a higher-power objective, which has a small depth of field. If possible, therefore, it is best to use the lowest magnification objective lens consistent with the nature of the sample being tested.

Appendix 2 indicates the relationship between reflectance and the micro-hardness (VHN) of the minerals described in the book.

80 Cooperite

Chemical Formula **PtS**	λnm	Air			Oil		
		R_o	R_e'	R	R_o	R_e'	R
Symmetry **Tetragonal**	470	42.5	48.4		26.6	33.3	
Provenance **Potgietersrus, S. Africa** 24°15′S 28°55′E	546	39.5	47.05		23.35	31.5	
	589	37.55	45.85		21.8	30.2	
	650	35.8	44.4		20.3	28.8	
Standard **WTiC (Zeiss, 314)**	400	40.9	46.6		24.5	30.5	
Monochromator Δλ≈12nm **Line interference filter**	420	41.95	47.3		25.6	31.5	
Photomultiplier **S 20, Hamamatsu type R 928**	440	42.6	47.9		26.5	32.4	
	460	42.7	48.3		26.7	33.1	
Effective N.A. **(plane glass reflector)** Air 0.35, 0.2 Oil 0.4, 0.2	480	42.2	48.4		26.3	33.4	
Chemical Composition **Pt 84.1**	500	41.5	48.2		25.5	33.0	
Pd 1.5	520	40.4	47.8		24.6	32.4	
Ni 0.8	540	39.4	47.2		23.6	31.7	
S 13.9	560	38.5	46.6		22.7	31.05	
100.3 wt%	580	37.85	46.1		22.0	30.45	
electron microprobe at 20kV with pure Pt, Pd, Ni standards and troilite	600	37.2	45.5		21.5	29.9	
X-ray Data: **Diagram corresponds to PDF No. 18–972**	620	36.6	45.1		21.0	29.45	
	640	36.0	44.7		20.5	29.1	
	660	35.6	44.2		20.1	28.6	
	680	35.1	43.7		19.7	28.2	
	700	34.6	43.3		19.35	27.8	

Colour values

		Air			Oil			Illum- inant
		R_o	R_e'	R	R_o	R_e'	R	
x		.298	.305		.291	.301		
y		.307	.314		.302	.313		
Y%		38.9	46.7		23.1	31.2		C
$λ_d$		481	486		481	486		
P_e%		5.5	2.3		8.6	3.8		
x		.435	.441		.428	.437		
y		.406	.408		.404	.408		
Y%		38.3	46.3		22.5	30.8		A
$λ_d$		492	497		492	496		
P_e%		3.0	1.5		4.8	2.4		

VHN: 762–920 (10 indentations)   Load (gf) 100
f-scc

Polishing Method:
800/1000 mesh carborundum on glass, 6, 3, 1 & ¼ μm diamond on Hyprocel laps on Engis machines, final buffing with MgO in distilled water on cloth

Reference & Further Information:
Criddle A.J., & Stanley C.J. (1985) Characteristic optical data for cooperite, braggite and vysotskite. *Can. Mineral.* **23** 149–162

Zeiss oil DIN 58.884 at 21°C

Specimen: E.700   BM 1932,1301

Fig. 4.2 Example of a card from the IMA/COM Quantitative Data File (Second Issue, 1986) showing general data for cooperite. Reproduced with permission of the British Museum (Natural History).

4.3   SIMPLE PHYSICAL TECHNIQUES

4.3.1   INTRODUCTION

Although this book is primarily concerned with the examination of samples which have been cut, mounted and polished, the student should whenever possible examine the hand sample from which the polished block has been prepared. A visual examination with a small hand lens (×8 or ×10 magnification) will some-times provide as much information as hours spent looking down a microscope. The following aspects may therefore be noted in either megascopic or microscopic fields of view, and the Mineral Tables (Ch. 8) give not only detailed information with respect to observations and tests conducted with the aid of the microscope and its attachments, but also the colour, streak and lustre of the sample as viewed by the unaided eye. These features may more easily be noted on a large hand sample, rather than with material which is present as minute grains. It is better to use the microscope and shine the light onto the surface of the sample from an external light source, e.g. a tungsten light bulb, excluding both polars. Normally a hand-held light is sufficient for this quick examination. It is very useful with non-opaque and the poorly reflective minerals, i.e. the gangue minerals.

The commonly examined physical properties are:

Crystal morphology
Zoning
Cleavage and parting
Twinning
Inclusions and intergrowths
Hardness
Tenacity
Streak

A number of these properties are also included in Chapter 5 describing the textures of the ore minerals.

4.3.2 CRYSTAL MORPHOLOGY (FORMS AND HABITS)

As with hand samples, in which the crystal forms and habit can be described, so too with the ore minerals in polished section, similar physical properties can be seen and described. The terminology may be the same though synonyms are sometimes used. Some minerals, frequently the harder ones, occur as well-formed crystals and the descriptive terms euhedral or idiomorphic or panidiomorphic may be applied as, for example, to pyrite, hematite, arsenopyrite, cobaltite, magnetite and wolframite. Other minerals may develop relatively few crystal faces and the terms subhedral or hypidiomorphic are applied. The third and largest group, often comprising the majority of the softer minerals, do not normally form crystal faces except where they grow in an open space. The terms anhedral or xeno-morphic or allotriomorphic may be used to describe these minerals.

The more general descriptive terms which have become standard in mineralogy may be used to describe the appearances of crystals, or crystal aggregates. A wide range of possible terms can be applied of which the following are typical examples:

Acicular	Dendritic	Radiating
Bladed	Foliate	Rounded
Botryoidal	Fibrous	Rhombic
Cubic	Lamellar	Skeletal
Columnar	Micaceous	Tabular
Colloform	Prismatic	Wavy

It is often difficult to be precise about the use of these terms.
Note: It should always be remembered that the outlines shown on a surface depend on the way that the surface intersects the crystal. A cube, for example,

may appear in cross-section as a rectangle, square or triangle. A considerable search may need to be made before a complete form profile is observed, or it may be that one does not exist at all in the sample.

4.3.3   ZONING

Zonal structure may be seen in ordinary or polarised light or it may require the sample to be etched before it becomes apparent. It appears as concentric (internal) bands parallel to the crystal faces or indicating the original crystal form. Zoning may be due to:

1. Depositional pauses during growth.
2. Different growth rates with or without inclusions.
3. Chemical variations (often minor) during growth.

Any mineral is capable of showing zoning but it is often found in arsenopyrite, galena, sphalerite, pyrite, stibnite and cobaltite.

4.3.4   CLEAVAGE AND PARTING

Cleavages and partings are readily seen in hand samples and thin sections observed in transmitted light, see Table 4.7.

There are four possibilities which may be seen:

1. Poorly polished samples.
2. Etched samples – unfortunately the whole surface often needs to be etched.
3. Weathered or altered samples. Decomposition, if it occurs, may take place along cleavage planes and partings but also along grain boundaries. Often in this case, it is better developed at the margins of grains as, for example, in the alteration of:

> Galena to cerussite or anglesite
> Chalcopyrite to covellite
> Sphalerite to smithsonite

Table 4.7   Typical cleavages exhibited by the ore minerals. Cleavages may occur in one or more directions

1. PERFECT	Bismuthinite
	Covellite
	Molybdenite
2. IMPERFECT	Galena
	Manganite
	Pentlandite
	Pyrolusite
	Stibnite
3. DISTINCT	Boulangerite
	Chalcocite

4. Polishing pits. These, if present, depend on the polishing technique and the orientation of the grains relative to the cleavage direction. They are frequently, but not invariably, seen in galena, magnetite, pentlandite and sphalerite.

Cleavage or parting may be seen as a series of parallel cracks or, if a number of cleavages are present, as rows of sub-parallel triangular pits. Cubic, octahedral and rhombohedral cleavages are shown in this way. Diamond-shaped, triangular or rectangular patterns result from prismatic cleavages and parallel cracks from pinacoidal cleavage.

Note: Cleavage is not often well developed in fine-grained samples, and even if a mineral is known to possess cleavage, it need not be apparent at all in a polished sample. As an example, enargite has four cleavages, but only one is seen, and rarely at that, in a polished sample.

4.3.5 TWINNING (SEE TABLE 4.8)

Growth twins or mechanical twins are developed in some of the ore minerals. In cubic minerals, twins are noted by abrupt changes in the orientation of the cleavages or of the pattern of inclusions or zoning. These features may sometimes only be seen when the surface is etched. Twinning in anisotropic minerals may or may not be noticed in plane-polarised light; it is best seen when viewed under crossed polars, when marked contrasts in colour may occur.

Growth twinning may be simple or complex and occur on numerous 'twin laws'. Polysynthetic twinning is very characteristic and may, for example, differentiate chalcopyrite from gold (which does not show growth twins). The arrowhead twins of marcasite are also very common.

Mechanical twins are due to deformation of the ore and are often associated with partings which may be curved or sinuous. The mineral grains may be bent

Table 4.8 Twinning shown by some of the ore minerals

1. Simple (or lamellar) Twins.
   Cassiterite, marcasite.

2. Polysynthetic Twins (often in various directions).
   Hematite, ilmenite, rutile, millerite.

3. Polysynthetic Twins (often in only **one** direction).
   Argentite, chalcocite, marcasite, sylvanite.

4. Complex Twins.
   Cobaltite, arsenopyrite.

5. Sinuous Twins.
   Stibnite, bismuthinite, cinnabar, pyrrhotite.

6. Star Twins (six-pointed).
   Löllingite, arsenopyrite.

7. Interpenetrant Twins (four-fold).
   Marcasite, cubanite.

8. Polysynthetic Twins (orthogonal).
   Stannite, bornite, chalcopyrite.

or fractured and give rise to undulose extinction patterns. Hematite and chal-copyrite may show deformation twins which appear as lamellae, while stannite often develops inversion twins. Stibnite, molybdenite, covellite and pyrrhotite can show very striking twin patterns.

4.3.6   INCLUSIONS AND INTERGROWTHS

These may occur on any scale but they are normally microscopical or submi-croscopical features. They occur singly or in groups and may be regular or irregular in outline. Their presence depends on the mode of formation of the host mineral and the inclusion. They may be subdivided into:

1.   Inclusions accidentally trapped during the growth of the mineral.
2.   Inclusions which illustrate zonal growth (see, Zoning, 4.3.3).
3.   Relics or remnants of a pre-existing mineral phase which has been replaced. They may be randomly or crystallographically orientated.
4.   Exsolution intergrowths or inclusions.
5.   Simultaneous deposition from two mineral phases.

For more detailed information on some of these features, see Chapter 5 on the textures of the ore minerals.

4.3.7   TENACITY

Tenacity describes whether a mineral is sectile, ductile or brittle. Most ore minerals are brittle and when scratched (see hardness section, 4.4) a powder results. However, a small number may, when scratched, produce a groove and be said to be ductile, or produce a very deep scratch which has some of the mineral raised at either side (not powder in this case). This latter group is said to be sectile.

  This test can distinguish only a small number of minerals as, for example, gold from chalcopyrite.

4.3.8   STREAK

The colour of a mineral is the streak, which is produced by scratching the surface and, as a result, producing a small amount of powder. Although not as useful as in hand-specimen mineralogy, it should be noted in ore microscopy using a low-power objective lens. Alternatively, shine a light obliquely at the face of the polished sample. Most streaks are black or grey.

4.4   HARDNESS

There are three types of hardness:

1.   Polishing hardness.
2.   Scratch hardness.
3.   Microindentation hardness.

The first two are relative and qualitative; the third can be measured more quantitatively.

4.4.1 POLISHING HARDNESS

This is a mineral's resistance to abrasion and as such is a useful property for the identification of the various components in a multimineral granular aggregate. The harder minerals being more resistant to polishing stand out slightly higher than the softer minerals and these *slight* differences are noted in the relative relief.

To test for relative relief, focus on a grain boundary and *raise* the microscope's ocular tube. A thin line of light, called the Kalb light line, will appear to *move in* towards the softer mineral.

Invariably when a sample is polished with the finest diamond paste and with extreme care, scratches may still be seen, especially so if the sample is a composite of minerals of highly varying hardnesses. As such, a polishing scratch may be seen traversing one grain, stopping next to another and then (possibly) continuing across an adjacent grain, if it is of the same composition or hardness as the first grain. A relative hardness is therefore easy to note in such a case.

4.4.2 SCRATCH HARDNESS

This is a simple test, if performed correctly. A sharp needle in a wooden handle is dragged across the face of the sample. No pressure at all need be applied to force the needle down, for its weight and the handle alone are sufficient to mark most samples. At first it might be possible to distinguish only hard and soft minerals by this method (Short, 1940) but, with time and practice, subtle variations in the relative hardness of a large number of minerals can be detected.

Possibly the best known scheme for determining scratch hardness is that of S.B. Talmage. He formulated seven degrees of hardness, each one of which is characterised by a specific mineral (cf. Mohs' Scale of Hardness):

A. Argentite   E. Niccolite
B. Galena   F. Magnetite
C. Chalcopyrite   G. Ilmenite
D. Tetrahedrite

The Talmage (1925) technique has not become standard practice but his hardness scale is useful as a relative guide.

Note: The hardness of a mineral may vary considerably with its orientation (e.g. stibnite). Also, with the harder minerals, care should be taken that a scratch is not confused with a fracture.

4.4.3 MICROINDENTATION HARDNESS

This is a quantitative technique which measures the resistance of a mineral to indentation. Specialised instruments (Fig. 4.3) are sold in order to perform this investigation. Basically, an indentation is produced by bringing a pointed diamond in contact with a mineral grain and then applying a known load to that diamond. As the hardness is a vectorial property, even in cubic minerals, a

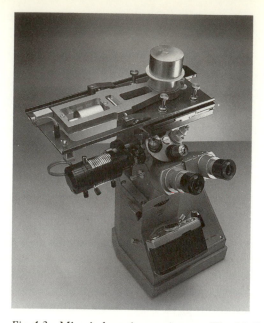

Fig. 4.3 Microindentation equipment. The McCrone low-load hardness tester mounted on an Olympus MG microscope. Reproduced by courtesy of McCrone Research Associates Ltd. and Olympus Optical Co. (UK) Ltd.

Table 4.9 Vickers Microhardness Values of the ore minerals (means and ranges)

Mineral	Vickers Microhardness	
	Mean	*Range*
Argentite	42	20–91
Arsenopyrite	1010	890–1283
Bismuthinite	127	92–172
Bornite	103	84–124
Boulangerite	148	116–217
Cassiterite	1201	992–1491
Chalcocite	74	58–98
Chalcopyrite	203	174–245
Chromite	1305	1036–1600
Cinnabar	77	61–99
Cobaltite	1179	948–1367
Copper	92	48–143
Covellite	82	68–110
Cubanite	218	150–260
Cuprite	204	179–249
Enargite	192	133–358
Galena	79	56–116
Goethite	659	525–824
Gold	65	41–102
Graphite	9	4–12
Hematite	948	739–1097

Table 4.9 (cont'd)

Mineral	Vickers Microhardness	
	Mean	Range
Ilmenite	629	519–739
Löllingite	742	421–963
Magnetite	581	480–734
Manganite	565	367–766
Marcasite	1121	824–1288
Millerite	281	192–383
Molybdenite	32	16–74
Niccolite	419	308–533
Pentlandite	220	195–238
Platinum	128	144–146
Proustite	119	90–143
Psilomelane	667	503–813
Pyrargyrite	100	50–127
Pyrite	1377	1027–1836
Pyrolusite	266	225–405
Pyrrhotite	278	212–363
Rutile	1046	978–1280
Scheelite	361	285–464
Silver	58	39–118
Sphalerite	199	128–276
Stannite	216	171–307
Stibnite	98	42–163
Sylvanite	144	102–203
Tennantite	358	308–401
Tetrahedrite	360	291–464
Uraninite	765	280–839
Wolframite	436	285–657

The data have been reproduced from McLeod and Chamberlain (1968) with the permission of the publishers: Geological Survey of Canada, Department of Energy, Mines and Resources.

considerable range of hardness values may be given by different grains of the same mineral.

In the Mineral Tables (Ch. 8) the Vickers Hardness is reported with a mean values as well as the range of values reported by numerous people. These figures are from McLeod and Chamberlain's (1968) comprehensive data tables, of which a partial listing is given in Table 4.9.

Appendix 2 also provides a chart which illustrates the relationship between microhardness (VHN) and the reflectance of the minerals described in this book.

5 TEXTURES OF THE ORE MINERALS

Abbreviations used

Barite .... Ba	Gold .... Au	Quartz .... Qtz
Bornite .... B	Gypsum .... Gyp	Sphalerite .... Sph
Chalcopyrite Cc	Ilmenite I	Stannite .... St
Chalcocite .... Chc	Magnetite .... M	Stibnite .... Sti
Chromite .... Cr	Pentlandite .... Pe	Tetrahedrite .... Te
Fluorite .... F	Pyrite .... Py	Wolframite .... W
Galena .... G	Pyrrhotite .... Pr	

5.1 INTRODUCTION

In reflected light microscopy and the study of the ore minerals especially, it is possibly as important to recognise and interpret the relationships between individual grains (i.e. the textures) as it is to identify the minerals.

Textures may assist in:

1. Indicating the nature of the processes of initial ore deposition.
2. Indicating subsequent 'events', such as re-equilibration, metamorphism, annealing, weathering, oxidation and cementation.
3. The recognition of extra-terrestrial textures (e.g. widmanstätten texture in iron meteorites).

The textural investigation of an ore can be very important in the milling and beneficiation of that ore into its component parts.

Certain minerals (oxides, sulphides and arsenides) are more likely to preserve their original textures than such minerals as pyrrhotite and the Cu–Fe sulphides. Other minerals are so prone to alteration (e.g. argentite, and the native metals) that it is rare to find them illustrating any of their original textures. Monomineralic ores are less prone to subsequent alteration than are the polymineralic ores which may undergo numerous superimpositions of textural change.

No scheme of textural classification is perfect and no scheme will ever be universally agreed. Certain terms have become 'standard nomenclature' in particular parts of the world. Hence you find certain 'schools of thought', i.e. the American, the European and the Far Eastern (Asia and Australia) where textural terms are used somewhat differently. Synonyms are frequently used and can be misleading. Textures can be described as zoned, concentric, rhythmic, colloform and crustified to imply a genetic origin, or purely to denote the visual appearance of a mineral or an admixture of minerals.

5.2 DESCRIPTIVE TEXTURES

5.2.1 SINGLE GRAINS

5.2.1(a) Internal
5.2.1(a1) *Zoning*
Etching may or may not be required in order to show zoning. The mineral may be colour or hardness zoned. The zoning may have resulted from interruptions

in growth, changes in the physical characteristics of alternating bands or the presence of inclusions in some bands and not in others. It normally indicates either repeated and renewed growth or rapid growth and a low temperature of formation from an impure fluid. Zoning may have occurred due to secondary processes, e.g. regional or contact metamorphism, or it may have been destroyed by such effects.

Examples: copper, stibnite, pyrite, galena, cassiterite and pyrargyrite.

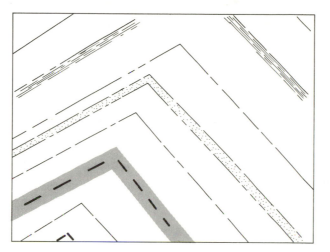

Fig. 5.1 Zoning.

Stoss-side growth is a form of zoning. It is revealed as a normal zonation pattern *but* bands on one or more sides of a crystal are wider than on other sides. A preferential deposition direction may be inferred, i.e. the fluid has come from the direction which has 'built up' the wider bands. It is commonly found in crystals growing in open cavities and vugs.

Examples: stibnite, millerite and many non-cubic minerals.

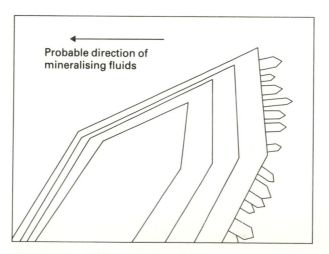

Fig. 5.2 Stoss-side zoning/growth.

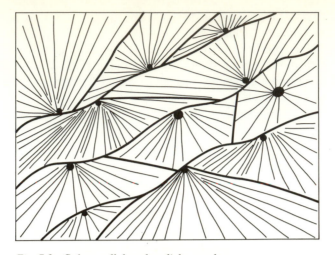

Fig. 5.3   Sub-parallel and radial growth.

Sub-parallel and radial growths are noted in minerals which naturally grow in open spaces and form columnar, prismatic or leaf-like forms. This may indicate a uniform supply of material from all directions, as opposed to stoss-side growth. Its precise form is related to the characteristic crystal growth pattern of the mineral concerned. If individual crystals compete with one another during growth, parallel or sub-parallel textures may result.

Examples: stibnite and most low-temperature minerals.

5.2.1(a2)   *Twinning*

Twinning may be classified by its inferred origin or by its visual appearance and the following types are recognised:

Growth twins. The lamellae of growth twins *are not* of a uniform thickness and may be interwoven and unevenly distributed throughout the sample. The texture is often dependent upon the temperature, kinetics and the degree of ore fluid supersaturation.

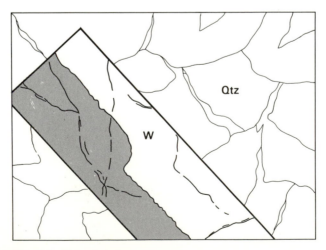

Fig. 5.4   Growth twins.

Examples: chalcopyrite, stannite, arsenopyrite, chalcocite and marcasite (arrowhead twins).

Defect and mosaic growth twins. Virtually every crystal contains *some* defects. They may be visible or invisible (microscopically). Natural etching (weathering, oxidation or tarnishing) or artificial chemical etching may reveal these defects. Microscopically they are better seen when the sample is near extinction under crossed polars.

Example: galena.

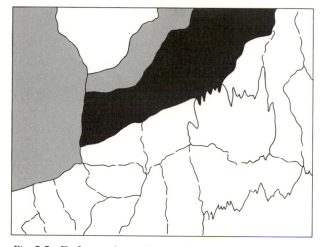

Fig. 5.5 Defect and mosaic growth twins.

Inversion twins. This is a rare phenomenon, but when it is developed it generally displays throughout the mineral grains a network of spindle-shaped intergrowths which are not parallel. It is even rarer to find it with strain and/or translation features. It indicates an initial formation above the inversion temperature and at least partial re-equilibration to the lower-temperature polymorph on cooling.

Examples: stannite, acanthite.

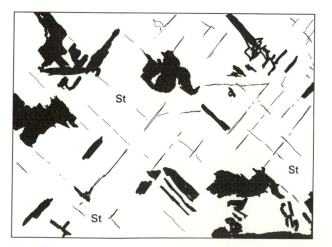

Fig. 5.6 Inversion twinning.

Pressure, lamellar or deformation twins. These are frequently noted and are often associated with ores which have undergone cataclasis, deformation (bending) or incipient recrystallisation. The lamellae are of uniform thickness, as opposed to growth twins, and tend to pass through adjacent grains.
Examples: hematite, chalcopyrite and many other minerals.

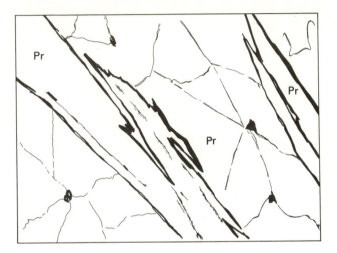

Fig. 5.7   Pressure, lamellar or deformation twins.

Crumpled lamellae. As Ramdohr (1980: 87) indicates, this texture has not yet been fully explained. It appears as spindle-shaped lamellae at right angles to bent or undulose translation planes.
Examples: graphite, molybdenite and covellite.

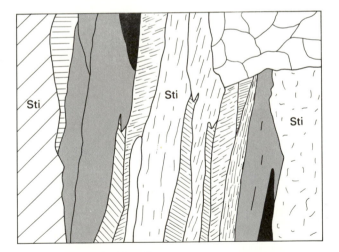

Fig. 5.8   Crumpled lamellae.

5.2.1(b)   External

5.2.1(b1)   Morphology

Xenomorphic (= allotriomorphic) texture. A term applied to a rock in which the crystals are anhedral. This texture is commonly noted in multimineralic aggregates due to the various rates of crystallisation.
Examples: silver, gold, galena, argentite and the fahlore group.

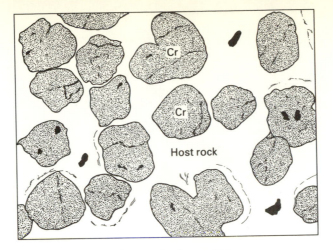

Fig. 5.9   Xenomorphic (=allotriomorphic) texture.

Hypidiomorphic texture. A term applied to a rock in which all the crystals are subhedral. This is not an uncommon texture and especially so when one of the components forms first *or* grows at a differential rate of crystallisation (cf. porphyritic and porphyroblastic textures in metamorphic terrains).
Examples: pyrite and garnets.

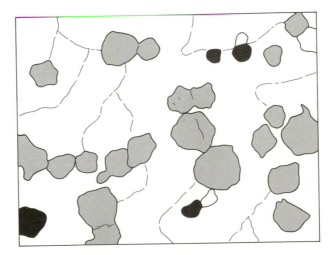

Fig. 5.10   Hypidiomorphic texture.

Idiomorphic (= panidiomorphic) texture. This is a very common texture describing a rock in which the crystals are euhedral and occurring mainly when growth has taken place in open cavities when either a continual or intermittent supply of fluid has been available. It is virtually impossible for this texture to form in monomineralic aggregates.
Examples: pyrite, arsenopyrite, cobaltite, löllingite, molybdenite, graphite and magnetite (rarely).

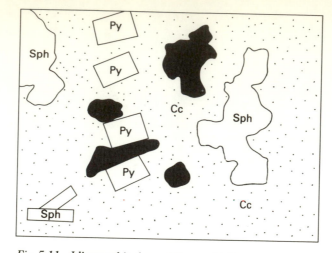

Fig. 5.11   Idiomorphic (=panidiomorphic)) texture.

5.2.1(b2)   *Grain size*

Size is a very subjective descriptive term, but one commonly used in petrography. Confusing terminology may result from the use of such terms as 'phenocrysts' which *only* implies that an idiomorphic grain occurs in a fine-grained groundmass. An 'idioblast', likewise, *only* implies an age relationship in that it is younger than the mineral adjacent to the idioblast. It may form at the 'expense', in space, of the other minerals.

Descriptive terms (i.e. coarse, medium, fine) are of limited use. A photomicrograph with a scale is essential or a full statement of the grain size in mm etc.

5.2.1(b3)   *Grain and crystal relationships*

The term 'bonding' indicates how the individual grains are held together, and it is frequently applied to monomineralic aggregates.

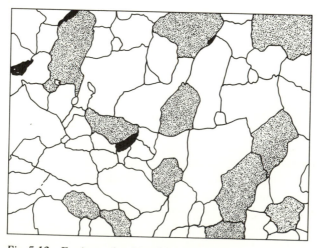

Fig. 5.12   Equigranular (simple) bonding.

Simple bonding. The simplest texture is equigranular which indicates a slow growth with no preferential direction. Smooth inter-grain contacts are characteristic.

Examples: many minerals.

Complex bonding. This descriptive term can cover virtually all crystal bonding textures and gives rise to numerous textural sub-types, e.g. interlaced, myrmekitic, etc. (see section 5.4.2).

Examples: many minerals.

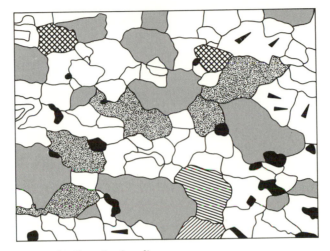

Fig. 5.13   Complex bonding.

Space filling. A compact specimen, in which there are no free or open spaces, is very rare indeed. Often it includes features which enable it to be called porous, vesicular, cellular, drusy, etc. The sizes and shapes of mineral grains can be varied.

Examples: many minerals.

Fig. 5.14   Space filling.

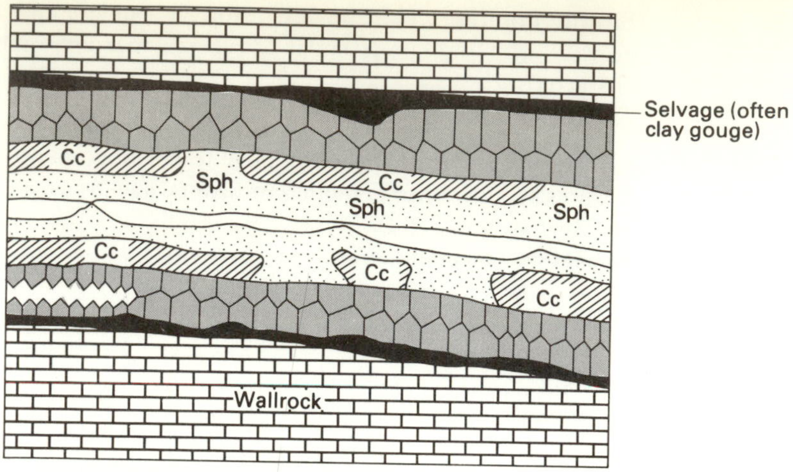

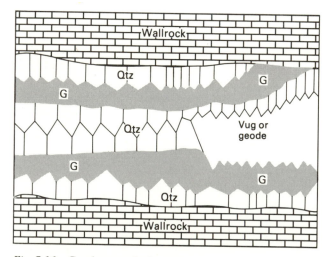

Fig. 5.15 Crustification banding.

Crustification banding. A texture which results from the successive deposition of minerals inwards from open fractures or fissure walls. It may be multimineralic or rhythmic, showing different (paragenetically) depositional events.
Examples: many minerals, but predominantly those of low-temperature hydro-thermal deposits.

Comb or cockade texture. Comb or cockade textures result from growth of crystals outwards from the walls of open fractures, i.e. a primary open-space depositional texture. The crystallinity may be xenomorphic, hypidiomorphic or idiomorphic. It is named from its visual resemblance to the teeth of a comb or a cockerel's comb.
Examples: many minerals deposited in open spaces or opening fractures.

Fig. 5.16 Comb or cockade texture.

5.2.1(c) Nucleation

In order to initiate the growth of a crystal, the structure must develop from a nucleus of some kind. If a nucleus is already present, it is called a seed.

Generally, the nuclei are formed at the initial precipitation of the crystal. When conditions are suitable for a solid to exist, crystallisation will occur, and not before. At this point (the nucleation point), the initial crystal structure arrangement is determined and formed from the minimum number of atoms or molecules.

A nucleus must grow rapidly and reach its critical point, after which it can survive and grow, or else it will go into solution again. The critical size is dependent on the degree of supersaturation of a solution, the amount of under-cooling in a melt or, in the vapour phase, on the size of the droplets.

If the nucleus is a stable seed of another material, that material must have a structure that is similar in some way to the nascent crystal. The new crystal is in fact an orientated overgrowth. The process is called epitaxis and the growth is called epitaxial growth. Such relationships exist for the nucleation of galena by rock salt and ice by covellite.

The density of nucleation points will influence the final product, so much so that oversaturation and the formation of numerous nuclei may result in a colloid or a gel. If fewer nuclei are formed, a fine crystalline aggregate results, and with a minimal supersaturation, alternating nucleation may be so infrequent that only a few large crystals are formed.

5.2.1(d) Intergrowths

5.2.1(d1) Emulsoid intergrowth

This texture may resemble exsolution texture (p. 61). It takes the form of one mineral finely disseminated in another more abundant mineral. The term implies an *even* distribution of *equal and small* grains. A genetic interpretation can be inferred as the micro-component may have been trapped during crystallisation or represent incomplete replacement. Most authors use this term to indicate emulsion-like precipitates from a gel due to a decrease in temperature.
Examples: many minerals.

×50

—As phase

—As Sb phase

×200

Fig. 5.17 Emulsoid intergrowth texture.

5.2.1(d2) Myrmekitic texture

An interfingering texture (Barton 1978) which has been given other names (eutectic, cotectic, granophyric and graphic), and which is more important in

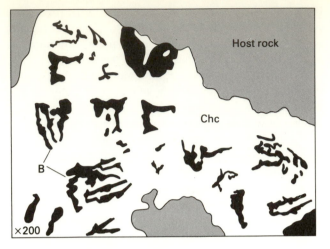

Fig. 5.18 Myrmekitic texture.

silicate rocks (MacKenzie *et al.* 1982). It appears as the growth of two or more minerals in variable amounts with mutually rounded boundaries, and has also been termed 'woven texture'. Depending on the angle at which the section has been cut, it may appear as lamellar intergrowths all with uniform optical orientation, or as a lattice-like texture.

Examples: a range of minerals have been reported showing this texture. For a comprehensive list see Ramdohr (1980: 111–17). A typical example is chalcopyrite–sphalerite (Barton 1978).

5.2.1(d3) *Orientated intergrowth*
A very common texture and one often related to the structures of the minerals. No specific genetic significance is indicated, although frequently it implies a higher than normal depositional temperature.
Examples: graphite–muscovite; stannite–chalcopyrite; pentlandite–pyrrhotite.

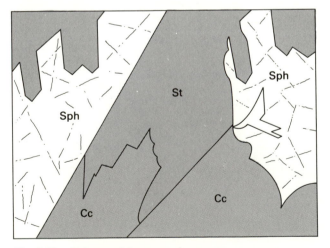

Fig. 5.19 Orientated intergrowth texture.

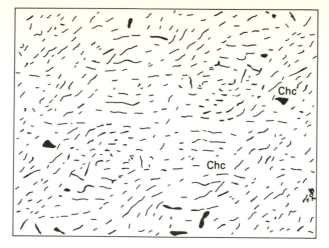

Fig. 5.20 Penetration intergrowths.

5.2.1(d4) *Penetration intergrowth*
A texture which is often visually similar to 'graphic granite' in petrography. It is *very* common in the ore minerals and occurs when any type of mutual penetration of two or more minerals occurs. A lamellar appearance often results and it *can* show transition to emulsoid intergrowths or caries-like textures.
Examples: many multimineral aggregates.

5.2.2   AGGREGATES

5.2.2(a)   Arrangement aggregates
5.2.2(a1) *Breccia aggregates*
Brecciated textures are very common and are frequently due to secondary tectonic processes, e.g. vein or fault reactivation of colloform, crustified or *any* other primary texture. Angular fragments of ores, gangue or wallrock lithologies may

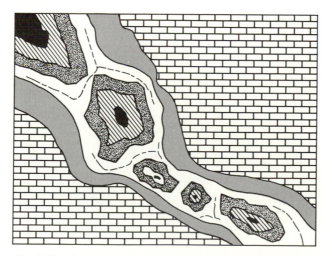

Fig. 5.21 Breccia texture.

all be introduced and subsequently overgrown or cemented. Size and shape can be on any scale.
Examples: any ore and any environment.

5.2.2(a2)   *Orientated, lineated or fibrous aggregates*
The orientation may be controlled by crystal shape or by crystal structure. In the former, roughly oval grains, for example, lie with the longer axes parallel to a preferred direction, perhaps due to an external influence, while in the latter case the orientation may be due to parallel growth, one crystal influencing its neighbours. It may also be termed **radial** or **bladed texture**.
Examples: some cubic minerals, quartz.

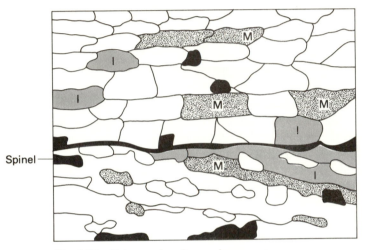

Fig. 5.22   Orientated or lineated aggregates.

A special case of orientated aggregates is **fibrous texture** in which an arrangement of parallel or near-parallel elongated units is present (see also comb or cockade). In sedimentary environments it may have been induced by shearing or it may be a primary growth feature.
Example: gypsum (satin spar).

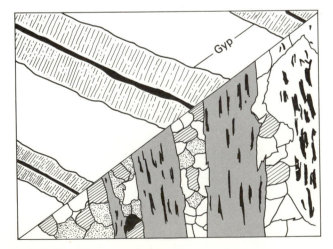

Fig. 5.23   Fibrous aggregates.

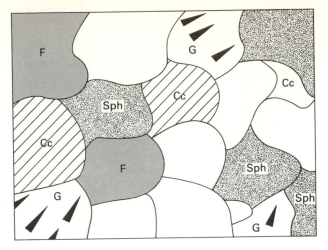

Fig. 5.24 Random aggregates.

5.2.2(a3) *Random aggregates*

A frequent term in petrography for equigranular undeformed rocks with mineral grains in no preferred orientation. Used also in describing ore minerals with such textures; *no* genetic significance.

Examples: all ore types, but especially in equigranular aggregates.

5.2.2(a4) *Rhythmic textures*

Rhythmic textures can be the result of numerous depositional processes of a primary or secondary origin (weathering or cementation), and as such are difficult, if not impossible, to relate to formation processes. Rhythmic growth may also be termed 'zoned', 'colloform' or 'crustified' texture. It is an essential feature of rhythmic textures that repeated mineral sequences are noted.

Examples: almost all environments and minerals.

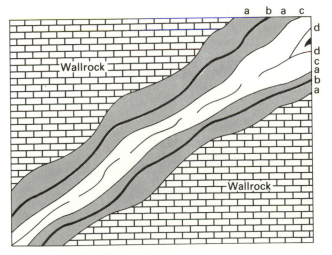

Fig. 5.25 Rhythmic textures.

5.2.2(b) **Contact and intergranular rims and films**

These textures, sometimes called reaction rims, may be due to crystallisation, exsolution, mineral 'migration' to grain boundaries, or true grain boundary

Fig. 5.26 Contact and intergranular rims and films.

reactions. A temperature difference may initiate or terminate such processes. The rims and films may vary in width and composition and may be due to primary or secondary interaction with the host grain(s).

Examples: (1) contact rims: crusts of bournonite, bournonite and the fahlore group, fahlore group and sphalerite ± bornite, with or without galena; (2) intergranular rims and films: virtually all minerals.

5.2.2(c) Inclusions

Inclusions of any mineral in any other mineral may occur. The minor mineral is termed the 'guest' and the major mineral the 'host' (not to be confused with exsolution texture terminology). The range of such inclusions is so large that it may be subdivided into at least four categories (Ramdohr 1980: 134):

1. Unrelated guests – primary and incorporated.
2. Related guests – orientated exsolution, reaction and devitrification.
3. Transformation products – thermal contact zones, heating/cooling environments, metamorphic and oxidation zones.
4. Immigrants – hypogene environments, cementation and deep weathering (which may include supergene) environments.

5.3   GENETIC TEXTURES DUE TO PRIMARY DEPOSITION

5.3.1   GROWTH FABRIC TEXTURES

5.3.1(a) Granular texture

Although commonly applied to magmatic and salt deposits, in ore deposits this textural term is infrequently used, since more definitive terms can be used, e.g. xenomorphic, hypidiomorphic and idiomorphic. Fine-grained textures may imply rapid crystallisation, and unequal granular textures may be due to differing rates of crystallisation.

Examples: all environments and minerals.

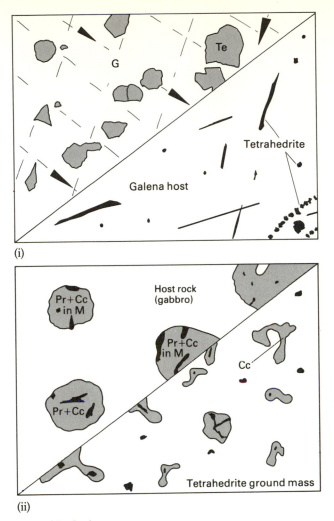

Fig. 5.27 Inclusions.

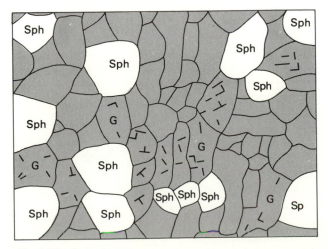

Fig. 5.28 Granular texture.

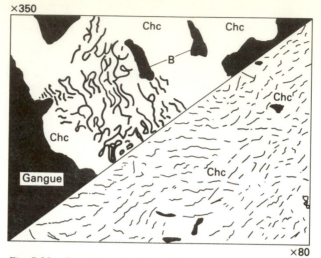

Fig. 5.29 Graphic and sub-graphic texture.

5.3.1(b) Graphic and sub-graphic texture

This is a common texture which may or may not be the result of eutectic crystallisation. The fabric often illustrated in petrography texts is of 'graphic' intergrowths so-called because of its resemblance to cuneiform writing. 'Sub-graphic texture' is used for *very fine* intergrowths of a similar nature, and as the distinction is somewhat subjective the latter term should be used with caution, or not at all. *Examples*:   all minerals and environments.

5.3.1(c) Zonal texture

Many minerals are zoned or banded. The textures may be restricted to one crystal (growth zoning) or involve a mass of crystals when it is called banding. Pauses in growth may result in compositional differences, crystallographic re-orientation or fractional crystallisation in a solid solution series. Concretionary zoning is found in oolites, magmatic zoning/banding may occur as a result of crystal settling or flowage, and crustification banding and zoning may develop from the deposition of minerals in a fracture, opening or fissure. Crustification banding and zoning may also exhibit 'comb or cockade' textures. The term 'Colloform' banding

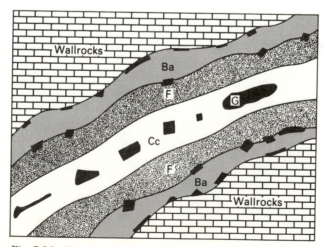

Fig. 5.30 Zonal texture.

and zoning, implying a colloid- or gel-texture, is used to describe minerals that occur in concentric curved or scalloped layers. If a free surface is present, the specimen's appearance may be termed reniform, botryoidal or mammillary. Edwards (1960) describes numerous minerals, of both a hypogene and supergene origin, which are capable of showing colloform zoning/banding.

Examples: native metals; hematite–ilmenite; wolframite–cassiterite; fahlore group–pyrargyrite.

5.3.1(d) Oolitic texture

A comprehensive text (Augustithis 1982) reports on this and spheroidal textures. This texture, usually zoned, may result from sedimentary processes, melts or solutions and can be mistaken for minute cockade or comb texture (Rieder 1969) if not examined carefully.

Examples: carbonate rocks as well as iron and manganese minerals.

Fig. 5.31 Oolitic texture.

5.3.1(e) Spheroidal texture

This texture is differentiated from oolitic texture by the *unequal* sizes of the spheroids and lack of zonal structures.

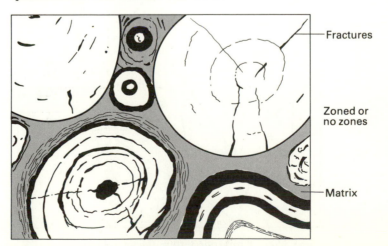

Fig. 5.32 Spheroidal texture.

Examples: pyrrhotite, bismuth minerals and mercury.

5.3.1(f)   Poikilitic texture

A texture resulting from the simultaneous crystallisation of two minerals where one is uniformly orientated and the other enclosed in sieve-like cells in the former. It is a rare texture and may be mistaken for some replacement textures.

Example: ilmenite in spinel.

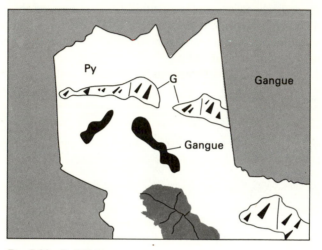

Fig. 5.33   Poikilitic texture.

5.3.1(g)   Porphyritic texture

Porphyritic texture results from direct precipitation from a melt and is therefore very rare in ore minerals. Pyrite, being deposited from a pure solution, may form this texture which consists of large iomorphic crystals embedded in a fine-grained groundmass or matrix.

Examples: pyrite, arsenopyrite and cobaltite.

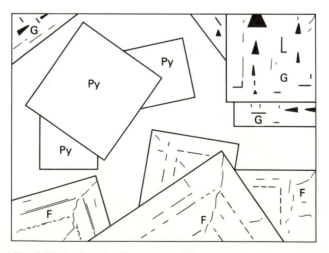

Fig. 5.34(a)   Porphyritic texture (idiomorphic).

Fig. 5.34(b) Porphyroblasts (porphyritic texture).

5.3.2 COLLOIDAL TEXTURES

The result of colloidal deposition, generally from solution, is not amorphous for colloidal precipitates are metastable with a tendency to crystallise. Microcrystalline aggregates are thus frequently observed due to the loss of the solvent (often water). Ramdohr (1980) indicates that transitional textures are possible and the individual forms can be different due to shrinking by dehydration, temperature and accidental effects as well as alteration with time. Commonly encountered textures are:

5.3.2(a) Botryoidal texture

The most common colloidal texture which, when etched, may indicate a more concentric structure than is normally visible. It may or may not contain inclusions. *Examples*: sphalerite (schalenblende) and numerous oxides.

Fig. 5.35 Botryoidal texture.

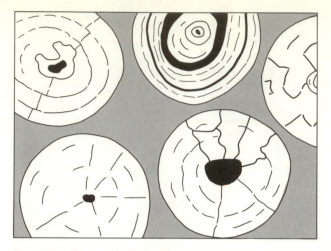

Fig. 5.36 Concentrically banded texture.

5.3.2(b)   Concentrically banded texture
A textural term which describes a visual appearance rather than a genetic origin.
A mineral may develop zonal texture which may be described as 'colloform' texture
or it may show a granular texture in a specific depositional layer. The layers may
form as discrete concentric shells or they may consist of fibrous or feathery crystals extending through more than one shell.
Examples: almost any ore mineral, but especially common in pyrite and arsenic,
antimony and arsenic–antimony solid solutions.

5.3.2(c)   Reticulate texture
This texture relates to colloidal processes as skeletal growth does to glasses. It
is often a precursor to botryoidal or concentrically banded texture and hence

Fig. 5.37 Reticulate texture.

appears at the nucleus of a colloidal mass. Extensive crystal nets of uniform orientation are typically seen.

Examples: galena and sphalerite (schalenblende).

5.3.2(d)   Spherical texture

Often termed 'bubbly' texture, it is the first formed of all the colloidal textures. In the germination of a gel, when a minute spheroid can serve as a nucleus, the subsequent overgrowths can be rhythmically concentric or grade into any of the other colloidal, zonal or banded textures. The spherical or bubbly name therefore describes a mass of these spherical nuclei.

Examples: all minerals capable of forming colloidal textures.

Fig. 5.38   Spherical texture.

5.3.3   SEDIMENTARY TEXTURES

True sedimentary processes give rise to a variety of genetically controlled textures, but some apparently sedimentary textures can be produced by hydrothermal or magmatic processes, in which case they should be termed pseudo-sedimentary or even epi-syngenetic textures.

In hydrothermal and magmatic environments, the intermittent deposition of an individual mineral, or crystal settling as the temperature or viscosity increases, decreases or alternates, may produce layered or banded textures. Metamorphic recrystallisation processes may also give rise to such features.

5.3.3(a)   Layered or banded texture

The texture is typified by the homogeneous rhythmic alternations in material, grain size and/or cohesion. The 'ore' may thus be of a varied mineralogy. The oolites in sedimentary sequences may be banded.

Examples: sands, sandstones, placers, beach sand deposits, banded iron formations (BIFs), magmatic ores (e.g. the Merensky Reef of S. Africa), metamorphic banding.

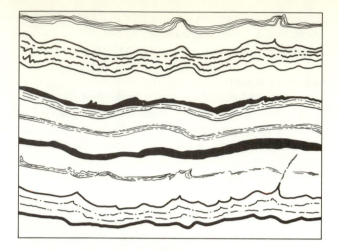

Fig. 5.39 Sedimentary layered or banded textures.

5.4 GENETIC TEXTURES RESULTING FROM TRANSFORMATION

A texture that indicates or implies that a primary mineral has been changed to its presently observed 'state' should be included in this classification. Thus paramorphic and polymorphic transformations, exsolutions, decomposition, replacement, cementation, oxidation, diagenesis, devitrification and leaching should be included. However, for simplicity this large group is subdivided into the specific groups described in the following text:

5.4.1 PARAMORPHIC REPLACEMENT TEXTURE

A texture which has resulted from the transformation of one mineral form of a mineral into another of the same mineral. The crystal morphology of a high-

Fig. 5.40 Pseudomorphic replacement texture.

temperature phase may be retained and it is sometimes not possible to recognise that the transformation has taken place.

Note: a paramorph is a pseudomorph which is also a polymorph, e.g. a calcite pseudomorph after aragonite or α-quartz after β-quartz.

Examples: marcasite to pyrite; only low-temperature paramorphs observed in chalcocite, acanthite and triolite.

5.4.2 EXSOLUTION – SIMPLE AND COMPLEX (MAY BE TERMED SECONDARY TEXTURES DUE TO COOLING)

True exsolution textures (Table 5.1) can only develop from minerals which are capable of forming solid solution series at high temperature. As the temperature falls, the solution becomes unstable and the individual components can 'exsolve' or 'unmix'. The terms 'decomposition' or 'eutectoidal breakdown' are also used to describe this phenomenon.

Table 5.1 Minerals which may show exsolution textures

Mineral	Exsolved mineral(s)
Argentite	Galena, bismuthinite, hessite*
Bismuthinite	Argentite
Bornite	Chalcopyrite, tetrahedrite–tennantite, chalcocite
Chalcocite	Bornite, chalcopyrite
Chalcopyrite	Bornite, pyrrhotite, millerite, cubanite, stannite, enargite, sphalerite, tetrahedrite–tennantite
Chromite	Hematite, ilmenite
Cinnabar	Tetrahedrite–tennantite
Copper	Gold
Cubanite	Chalcopyrite, stannite, sphalerite
Enargite	Chalcopyrite
Galena	Argentite
Gold	Copper
Hematite	Ilmenite, rutile, chromite
Ilmenite	Hematite, chromite, rutile, magnetite, gangue minerals, corundum* and spinels*
Magnetite	Ilmenite, pyrrhotite, garnets* and olivines*
Millerite	Chalcopyrite
Pentlandite	Pyrrhotite, chalcopyrite
Platinum	Iridium*
Pyrrhotite	Chalcopyrite, pentlandite, magnetite
Rutile	Ilmenite, hematite
Silver	Dyscrasite*
Sphalerite	Pyrrhotite, tetrahedrite–tennantite, stannite, cubanite, chalcopyrite
Stannite	Sphalerite, chalcopyrite, cubanite
Tetrahedrite/Tennantite†	Chalcopyrite, cinnabar, sphalerite, bornite

* Minerals not described in this book, but which may frequently occur.

† Tetrahedrite and tennantite are often known under the term 'fahlore'.

Exsolution bodies can vary in size from submicroscopic to pegmatitic, but in most instances the bodies are minute and will require a high magnification and possibly oil immersion study.

Exsolution textures (Table 5.2) can, and often do, resemble replacement textures which have quite a different origin.

Table 5.2 Common exsolution textures exhibited by the ore minerals

Host	Guest	Intergrowth pattern (with max. ratio of guest : host)
Argentite	Hessite*	Emulsion (1 : 5)
Bismuthinite	Argentite	————
Bornite	Chalcopyrite	Laths (1 : 3)
Bornite	Chalcocite	Network – cubic
Bornite	Tetrahedrite/Tennantite	———— (1 : 10)
Chalcocite	Chalcocite, Digenite*	Network
Chalcocite	Bornite	Network (1 : 2)
Chalcocite	Chalcopyrite	————
Chalcopyrite	Bornite	Laths – cross cutting
Chalcopyrite	Cubanite	Laths
Chalcopyrite	Sphalerite	Star/stellar/crosses
Chalcopyrite	Millerite	———— (1 : 6)
Chalcopyrite	Pyrrhotite	————
Chalcopyrite	Tetrahedrite/Tennantite	Emulsion (1 : 10)
Chromite	Hematite	Network/lamellae (1 : 5)
Chromite	Ilmenite	Lamellae (1 : 10)
Cubanite	Chalcopyrite	Irregular/Tabular
Enargite	Chalcopyrite	———— (1 : 10)
Galena	Argentite	Emulsion (1 : 10)
Gold	Copper	Lattice (1 : 3)
Hematite	Ilmenite	Lamellae/lens
Ilmenite	Magnetite	————
Ilmenite	Hematite	Lamellae/lens
Ilmenite	Rutile	Plates (1 : 10)
Magnetite	Ilmenite	Lamellae in a triangular pattern (1 : 1)
Magnetite	Ilmenite + (spinel)	Lamellae in a triangular pattern (1 : 5)
Pentlandite	Pyrrhotite	Emulsion (1 : 4)
Pentlandite	Chalcopyrite	Cell, cellular, netlike (1 : 10)
Platinum	Iridium*	Emulsion (1 : 4)
Pyrrhotite	Pentlandite	Flame/lamellae (1 : 5)
Pyrrhotite	Chalcopyrite	———— (1 : 4)
Pyrrhotite	Magnetite	Plates (1 : 10)
Rutile	Hematite	————
Rutile	Ilmenite	Lamellae (1 : 10)
Silver	Rare minerals*	Leaf-like
Sphalerite	Pyrrhotite	Blebs in rows
Sphalerite	Chalcopyrite	Emulsion (1 : 4)
Sphalerite	Chalcopyrite + Pyrrhotite + Cubanite	Emulsion (1 : 4)
Sphalerite	Tetrahedrite/Tennantite	Lamellae (1 : 5)

Table 5.2 (cont'd)

Host	Guest	Integrowth pattern (with max. ratio of guest : host)
Sphalerite	Stannite	Irregular blebs (1 : 10)
Stannite	Sphalerite	Emulsion (1 : 5)
Stannite	Chalcopyrite + Sphalerite	Emulsion (1 : 5)
Stannite	Chalcopyrite	Lamellae in a triangular pattern (1 : 2)
Stannite	Cubanite	Emulsion
Tetrahedrite/Tennantite	Chalcopyrite	Lamellae/network (1 : 5)
Tetrahedrite/Tennantite	Cinnabar	Emulsion (1 : 5)

Cassiterite may also show exsolution textures.

* Minerals not described in detail in this book.
Reproduced, with minor modifications, from Ramdohr (1980: 185–93) with the permission of the publishers (Pergamon Press).

It is usual to refer to the host (or parent) and guest (or daughter) phase. The host normally predominates and the guest occurs in a specific crystallographic orientation. The guest can increase in proportion so that it may predominate in a particular field of view of the sample being examined.

The following points may assist in the differentiation of exsolution textures from other textures which resemble them:

1. When blades, blebs or lamellae of the exsolved mineral cross each other, a distinct thinning of each bleb/blade occurs at the intersection. High magnification may be required in order to observe this effect.
2. If exsolution has resulted in the formation of both blebs and lamellae of the guest in the host, it is commonly noted that the number of blebs *decrease* in number and concentration as the lamellae are approached.
3. The presence of straight boundaries to the blades/lamellae, etc. often indicates that they were formed by exsolution.
4. Exsolution textures may lack any relationship between the exsolved phase(s) and the grain boundaries of the mineral in question.
5. Normally, exsolution phases occur as *discontinuous* units.
6. If the minerals under examination form a solid solution series, then it is most probable that somewhere on the specimen being examined, exsolution textures will have been developed.

Some of the minerals which are capable of showing exsolution textures are listed in Table 5.1 and Table 5.2 gives some of the most frequently noted exsolution textures between the minerals listed.

5.4.2(a) Crystallographic or orientated intergrowths
5.4.2(a1) *Cell (cellular, mesh, net or rim) exsolution texture*
In this texture the material which has exsolved lies at the edges of the host and thus appears to be filling the interstices. This texture may easily be mistaken for cementation, replacement or metamorphic textures.
Examples: bornite–chalcocite and pentlandite–chalcopyrite.

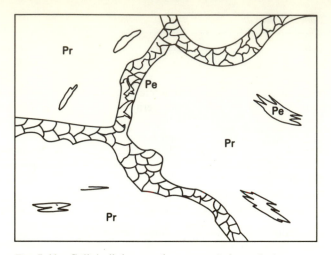

Fig. 5.41   Cell (cellular, mesh, net or rim) exsolution texture.

5.4.2(a2)   *Emulsoid or emulsion texture*

Rounded globules of the exsolved mineral (guest or daughter) normally occur at random in the host or parent. It may appear similar to spherical replacement texture, but in this case the guest would be noted at grain boundaries and as discrete grains.

Examples: galena–argentite,   tetrahedrite/tennantite–cinnabar,   stannite–cubanite, stannite–chalcopyrite, sphalerite–chalcopyrite and pentlandite–pyrrhotite.

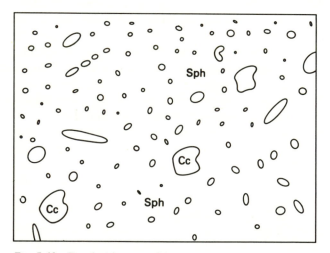

Fig. 5.42   Emulsoid or emulsion texture.

5.4.2(a3)   *Flame texture*

The name fully describes this exsolution texture which may form flames of the daughter totally in the parent or may be concentrated along the edges of secondary (cross-cutting) veinlets.

Examples: pentlandite (daughter) in pyrrhotite (parent).

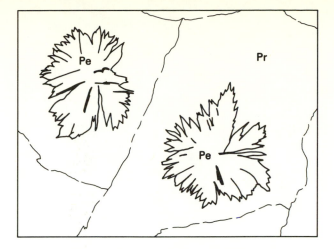

Fig. 5.43 Flame texture.

5.4.2(a4)   *Rosette, star or stellar texture*

More or less perfect star-shaped exsolution bodies of the daughter in the parent are observed. (Normally very high magnification is required in order to confirm that they are star-shaped). They are invariably crystallographically, not randomly, orientated.

Example: sphalerite in chalcopyrite.

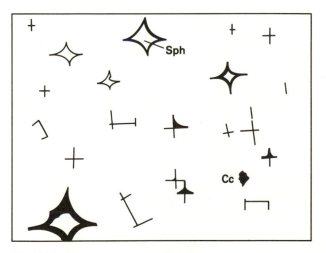

Fig. 5.44   Rosette, star or stellar texture.

5.4.2(a5)   *Seriate texture*

This texture is seen as blebs or blades of the exsolved body in the parent grain. Around the largest of the blebs, depletion of the exsolved material may be evident.

Examples: iron and titanium oxides.

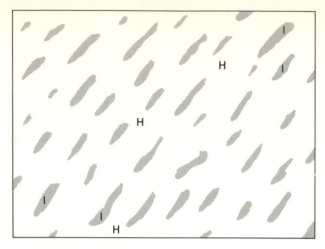

Fig. 5.45   Seriate texture.

5.4.2(a6)   *Widmanstätten (lattice, lamellar or triangular) texture*
Occurs as lath-shaped exsolution bodies orientated along the crystallographic directions of the parent crystal. They are often parallel to the octahedral planes of a cubic mineral.
Examples: often found in metallic meteorites which may require etching to reveal this texture. Magnetite–ilmenite.

Fig. 5.46   Widmanstätten (lattice, lamellar or triangular) texture.

5.4.3   REPLACEMENT TEXTURES

The genetic interpretation of these textures is very difficult. It requires years of experience and often intuition to ascertain or even approach the correct answer (Rieder 1969). Normally, replacement occurs along fractures, fissures, faults, grain boundaries and cleavage planes. Illustrations of the texture are therefore given rather than an attempt to explain the sometimes complex mechanism of formation.

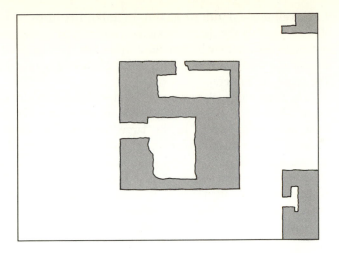

Fig. 5.47 Caries texture.

5.4.3(a) Caries texture

Named from the dental term for the corrosion of a tooth. In ore microscopy, it is used to indicate the initial corrosion of a grain. Replacement frequently starts at the corners or on the crystal edges or cleavage planes. Advanced alteration may lead to *skeletal replacement* (Fig. 5.55).

Example: sphalerite–pyrite.

5.4.3(b) Boundary or frontal texture

A very frequently observed texture, for if the mineral has no cleavages, fractures, grain boundaries or inclusions, i.e. lines of weakness, then boundary or frontal replacement can occur, and this results in rounded and smooth mutual grain boundaries. The effect can be produced by simultaneous deposition and not replacement, but the presence of relic(t) textures can prove that replacement has taken place. Without such relic(t)s, it is virtually impossible to prove whether or not replacement occurred.

Example: chalcocite–bornite.

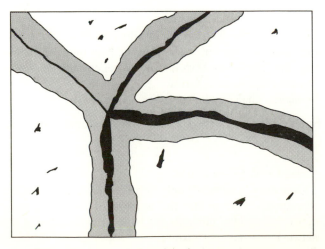

Fig. 5.48 Boundary or frontal replacement texture.

5.4.3(c)   Cell, cellular–island, or island–mainland texture

This is possibly the most commonly encountered replacement texture. However, a pseudo-texture may be produced by a cross-section of an interlocking granular mosaic. In this case, part of the polished face may reveal islands of the host in the replacing mineral.

The degree of replacement in island–mainland texture can vary from an initial to an advanced stage with all possible intermediates. Originally the fabric of the host may be preserved in the large 'islands', but as these are progressively reduced in size with advancing stages of replacement, the fabric of the whole is lost. Ramdohr (1980: 199) notes that the term *'caries-like'* has been used to describe the initial stages of island–mainland texture. As this texture (caries texture) is easily identifiable, it is considered in this text as a subdivision of island–mainland, *as well as* a variant of rim replacement.

Example: coal replaced by pyrite.

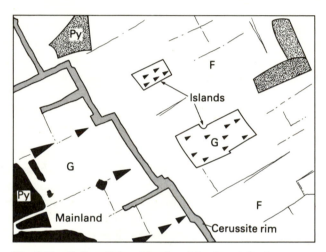

Fig. 5.49   Cell, cellular–island or island–mainland texture.

5.4.3(d)   Cleavage and rim texture

This texture results from the invading mineral 'eating' its way along grain boundaries and cleavages (if present) in the host mineral. In some circumstances it may resemble exsolution textures where penetration of the host has developed along crystallographic boundaries.

Example: chalcopyrite–covellite.

Subdivisions of cleavage and rim replacement texture are:

5.4.3(d1)   Network replacement texture

This texture is normally only formed when the host minerals have well developed cleavage(s). The greater the number of cleavages, the better developed is the network. Lamellar minerals and lamellar twins often, if replaced, yield an excellent network replacement texture.

Example: pyrite replaced by sphalerite along {100} cleavage.

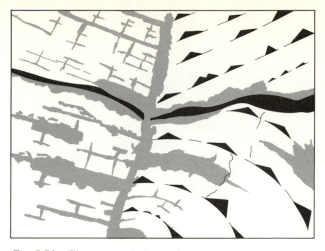

Fig. 5.50   Cleavage and rim replacement texture.

Fig. 5.51   Network replacement texture.

5.4.3(d2)   *Dendritic replacement texture*

An unusual texture (and not the same as dendritic deposition as, for example, of Mn ores on limestone surfaces), which develops due to replacement along three cleavages or partings or twin lamellae. It is a more elaborate form of network texture.

Example: calcite replaced by galena along twin lamellae.

5.4.3(e)   Core and zonal replacement texture

As the name implies, replacement of a zoned mineral may occur centrally (i.e. in the core of the host mineral) or selectively in one, some or all of the zones. As with other replacement textures, it is progressive and can vary from an initial and incipient replacement to an advanced form known as atoll texture.

Examples: **Core**: pyrite replaced by galena, sphalerite and chalcopyrite; galena replaced by anglesite; pitchblende replaced by sulphides and cobaltite replacing silver. **Zonal**: pyrite replaced by galena or sphalerite.

Fig. 5.52 Dendritic replacement texture.

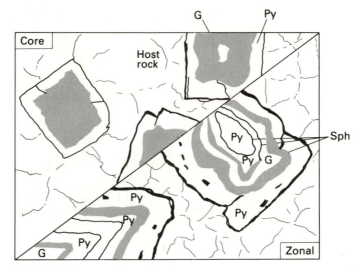

Fig. 5.53 Core and zonal replacement texture.

5.4.3(e) *Atoll texture*

This occurs when replacement of the core has subsequently progressed to virtually complete replacement of the remainder of the crystal. Only fragments of the host remain and these are frequently located along the crystal's boundary. *Example*: pyrite replaced by galena or chalcopyrite.

5.4.3(f) **Skeletal replacement texture**

In this texture (also called relic(t) texture), replacement has progressed to such a degree that only the last remnants of the original host mineral are visible. *Examples*: galena and pyrite.

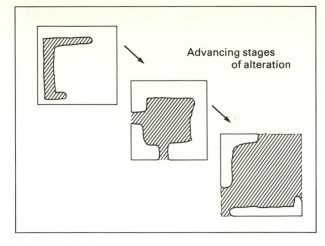

Fig. 5.54 Atoll texture.

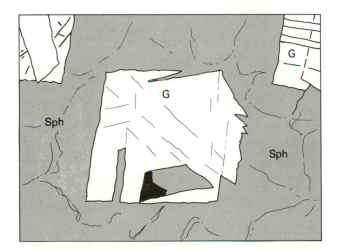

Fig. 5.55 Skeletal replacement texture.

5.4.3(g) **Vein and fracture replacement texture**
The criterion for recognition of vein replacement is that the opposite walls of the vein cannot be matched. If a filling had invaded an open fracture, then it would, in all probability, be possible to cross-correlate the opposite walls or margins.
Examples: gold and quartz in galena; gold in pyrite.

5.4.3(h) **Additional replacement textures**
The replacement of magnetite by hematite along {111} cleavages is called **martitization**. It may also be termed an oxidation texture.

The decomposition/alteration of pyrrhotite to a fine-grained mixture of pyrite and marcasite results in **bird's eye texture**.

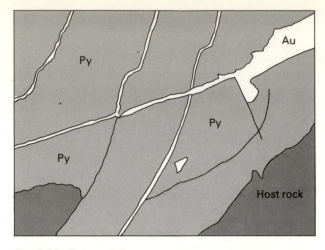

Fig. 5.56 Vein and fracture replacement texture.

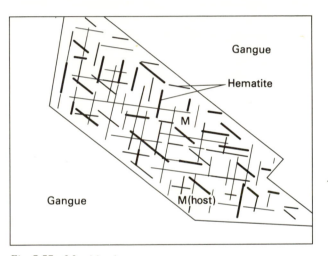

Fig. 5.57 Martitization texture.

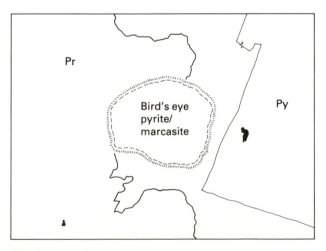

Fig. 5.58 Bird's eye texture.

5.4.4   RELIC(T) TEXTURES

A relic(t) texture is a pseudomorph of one mineral after another mineral and by its very nature is secondary and may assist in erecting a paragenetic sequence.

Every relic(t) texture *must* have originated at an earlier mineral stage, as some other texture. If the relic(t) transformation has been complete, it may be very difficult to deduce the former mineral phase. Zoning, exsolution, orientated inclusions and reticulate textures may all assist for recognition of a relic(t) texture. The term relict is preferred to relic.

5.4.4(a)   Replacement (relic(t)) textures

Pseudomorphs of one mineral after another are indicative of replacement, as is the preservation of original textures. The replacement, which has produced the relic(t), may occur with varying degrees of penetration. A mesh or lattice texture is often developed in minerals with strongly developed cleavage planes, while if the original mineral was zoned the relic(t) may exhibit zonal texture with either intermittent replacement or total alteration.

Examples: chalcopyrite to limonite; tetrahedrite to galena with or without bournonite; chalcopyrite after pyrite and covellite; and anglesite after galena.

Fig. 5.59   Replacement (relic(t)) texture.

5.4.5   DECOMPOSITION TEXTURES

This term is used by ore microscopists but it is impossible to distinguish decomposition textures from other textures and especially from exsolution textures. No further comments are therefore made on this very controversial and problematical subdivision of the textural classification of the ore minerals.

5.4.6   OXIDATION (WEATHERING) TEXTURES

Within this subdivision are some of the most important, diverse and complicated textures. Only a summary is given here and for further reading Edwards (1960), Rieder (1969), Ramdohr (1980), Craig and Vaughan (1981) and Picot and Johan (1982) are recommended.

A study of these textures (especially gossan textures) helps to infer whether the mineral assemblage is present in more or less the same state at greater depths or whether the enrichment is of a secondary origin, and is thus restricted to near-surface environments.

Descendant (supergene) fabrics may be distinguished from ascendant (hypogene, often misleadingly called primary) fabrics.

Oxidation can take place with or without volume changes and commonly cavities are formed which may be infilled with 'secondary' carbonates, sulphates, oxides or any of the soluble (transportable) components of the ore assemblage. The following list is not comprehensive:

Parent mineral	*Cavity infilling component*
Chalcopyrite	copper carbonates, limonite, chalcocite, covellite
Galena and other lead ores	cerussite, anglesite
Sphalerite	smithsonite
Pyrite	limonite

5.4.6(a)   Gossans

A gossan or 'iron-cap' is the oxidised outcropping cellular mass of limonitic material and gangue that overlies sulphide ore deposits. It has been formed by supergene leaching of the sulphide ore. A gossan may extend down as far as the ground-water table, and in some cases it may constitute ore.

Oxidation normally destroys all sulphides, selenides, tellurides and sulphosalts as well as arsenic and antimony salts. It may be rapid, resulting in total conversion, or it may proceed to partial or intermediate stages.

Pyrite can be very stable to weathering but in acidic environments it is one of the first minerals to dissociate completely.

Manganese ores all tend to change to pyrolusite and 'psilomelane' and often give rhythmic textures on alteration.

Textural fabrics exhibited by these ores are porous, irregular, laminate and colloform bands of various iron oxides with or without the presence of some of the residual sulphides. Relic(t) grains, cleavages and fractures may be preserved.

Fig. 5.60   Boxwork texture.

Granular (<10μm) or radiating fibrous textures may be developed. The open boxwork texture of cellular criss-cross laths of goethite and hematite, with or without pyrite, is a common texture.

Examples: worldwide in goethite, hematite and limonite with or without manganese oxides.

5.4.7 CEMENTATION TEXTURES

These textures normally occur only below ground-water level in the so-called zone of cementation. A typical example is the reaction of sulphate solutions on sulphide minerals, and cementation textures may be observed in: copper, silver, gold, platinum, mercury, lead and zinc minerals.

Precipitation frequently forms crusts with rhythmic and zonal textures with porous cavities. At an advanced stage it may leave only relic(t)s of the pre-existing minerals.

Cementation textures may be recognised by either/or of the following features:

1. Loose cellular masses.
2. 'Sooty' looking secondary minerals.
3. Rhythmic encrustations.

5.5 DEFORMATION TEXTURES

Any ore may show the effects of deformation, the degree of which will depend on the minerals present, the temperature and any other deformational processes.

A primary consideration as to whether a mineral phase is capable of being deformed is its hardness. A mineral which is susceptible to deformation is also prone to recrystallisation and so any deformation effects it may have suffered may be partially obliterated.

5.5.1 CURVATURE OF LINEAR FEATURES

Deformation of an ore may be indicated by the curvature or dislocation of crystal faces, lamellae, layers, veins or fractures. Deformation-induced twin lamellae may

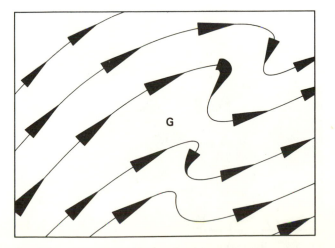

Fig. 5.61 Curvature of linear features.

be distinguished from growth or inversion lamellae since the former cut across grain boundaries whereas the latter do not. The deformation of exsolution intergrowths is not as easy to prove because the curvature of the laths, flames, rods or blebs of exsolved material may be intrinsic to the exsolution process.

Folding may produce ductile flow or brittle fracture depending upon the properties of the mineral concerned.

Examples: galena, pyrrhotite, ilmenite and chalcopyrite.

5.5.2   SCHLIEREN

A schlieren zone is a zone along which the ores are often pulverised and/or elongated through shearing and deformation. As a result planar or flattened lenticular features are produced in the ores.

5.5.3   BRECCIATION OR CATACLASIS

Often the more brittle components of an ore mineral assemblage (e.g. pyrite, chromite and magnetite) are more fractured and/or brecciated relative to the softer ore and gangue minerals.

Moderate deformation can grade into fragmentation and the ultimate stage is termed **durchbewegung** (Vokes 1969) when pulverisation of all the ore and gangue minerals gives rise to the so-called 'ball-textures', see Fig. 5.64.

5.6   ANNEALING TEXTURES

Annealing textures may be due to either the slow cooling or the slow heating of an ore.

5.6.1   RECRYSTALLISATION TEXTURES

Recrystallisation is the most frequently observed effect in annealing textures. It manifests itself by producing, in monomineralic aggregates, a mosaic of grains with

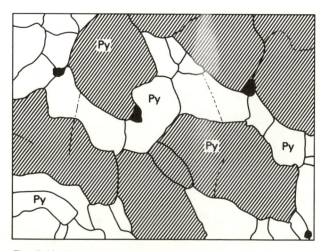

Fig. 5.62   Recrystallisation texture.

approximately 120° interfacial angles at their junctions. In polymineralic aggregates the interfacial angle depends on the minerals present (Stanton 1972).
Examples: pyrite, galena, sphalerite, chalcopyrite and pyrrhotite.

5.6.2   RE-EQUILIBRATION TEXTURES

5.6.2(a)   Overgrowths, homogenisation and porphyroblasts

Re-equilibration can result in overgrowths, homogenisation of the grains and porphyroblastic growth. Zoned overgrowths can result from annealing. Euhedral porphyroblasts (cf. garnets, pyrite and staurolite in metamorphic rocks) may also develop if the composition of the ore is suitable.
Examples: (1) overgrowths: pyrite on pyrite; (2) homogenisation: sphalerite, tethrahedrite; (3) porphyroblasts: pyrite, arsenopyrite, magnetite, hematite.

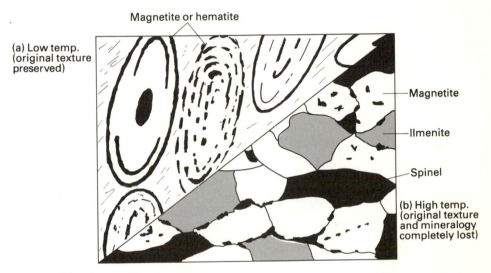

Fig. 5.63   Re-equilibration texture.

5.7   METAMORPHIC TEXTURES

For a comprehensive review of all metamorphic textures, Spry (1969) is recommended. Very few ore deposits have originated due to metamorphism but numerous deposits have been either mildly or totally altered as a result of it. Metamorphic effects are mainly due to pre-metamorphic mineralogy and the resulting textures relate to the original structures and the extent of either thermal or dynamic effects.

On a regional and microscopical scale, ores which have been affected are coarsened in grain size and may show drag and isoclinal folding or the attenuation of fold limbs, schistosity, rupturing, brecciation or boudinage.

5.7.1   DYNAMIC METAMORPHIC EFFECTS

These effects manifest themselves in the fracturing and/or recrystallisation of brittle ores or by producing ductile effects in the softer ores. A combination effect

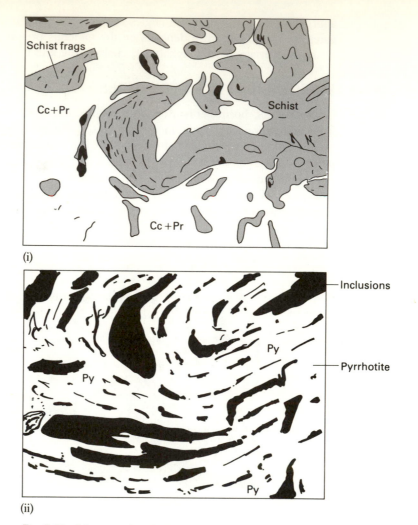

(i)

(ii)

Fig. 5.64　Metamorphic 'ball' texture.

may occur when the ductile ores infill fractures in the brittle ores. Deformed twinning, curved cleavage traces, kink-banding, undulose extinction and curved rows of crystallographically orientated inclusions may be observed. If intense deformation has resulted, **all** pre-metamorphic textures are lost and chaotic textures result, with disaggregation of primary banding and tectonic incorporation in the wallrocks (see Vokes 1969 and 1973). Ductile sulphides may be forced along cleavage planes or into relatively low-pressure areas. In these circumstances a paragenetic sequence cannot be erected.

Examples:　galena, chalcopyrite, sphalerite, pyrrhotite and pyrite (some may require etching to show the texture).

5.7.2　THERMAL METAMORPHIC EFFECTS

The ores may have already undergone dynamic metamorphism, but the effects may again result in the coarsening of grains, 120° triple junctions in monomineralic ores and recrystallisation in polymineralic ores with the result that small lens-

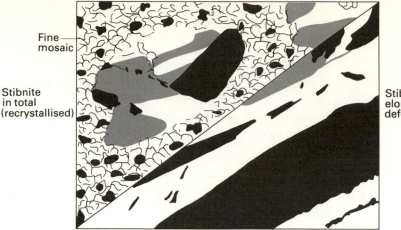

Fine mosaic

Stibnite in total (recrystallised)

Stibnite (in total) elongated and deformed

Fig. 5.65 Thermal metamorphic textures.

like grains occur along some of the original grain boundaries. Porphyroblastic growths or overgrowths are also the result. The range of textures exhibited will depend on the grade of metamorphism. At low grades, the more refractory minerals will not be affected and the low-temperature ones will be recrystallised. With increasing metamorphism, some of the more resistant minerals may lose some of their more mobile components (sulphur), be oxidised and/or totally recrystallised into a pseudomorphic form.

Examples: (1) Low grade: pyrite to porphyroblasts with growth zones; chalcopyrite and sphalerite produce equant anhedral forms; galena recrystallises; (2) Higher grades: pyrite goes to pyrrhotite and then oxidises to magnetite; sphalerite and tetrahedrite may lose their zoning.

5.7.3 SKARNS

Skarns or tactites are formed in very high-temperature contact metamorphic aureoles of intrusions into principally carbonate and less commonly aluminous and

Fig. 5.66 Skarn textures.

silicate-rich rocks. Coarse-grained and zonal textures predominate in iron oxides and sulphides. Disseminated grains and veinlets of sulphides, molybdates and tungstates are also noted. The carbonate hosts often contain manganese silicates, and replacement textures of carbonates and silicates are common.

Compositional banding, on all scales, as well as lamination of grains results. Secondary, reaction or deposition rims may be seen in some of the more 'mobile' ores.

Examples: wolframite, scheelite, magnetite, pyrite, chalcopyrite, bornite, covellite and the nickel and cobalt ores.

5.8   MISCELLANEOUS OR SPECIAL TEXTURES

A number of ore textures either cannot be classified into any of the subdivisions suggested, or are so characteristic that they warrant special names. One such texture is described below.

5.8.1   FRAMBOIDS OR FRAMBOIDAL TEXTURE

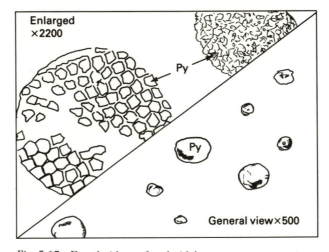

Fig. 5.67   Framboids or framboidal texture.

These are aggregates of spherical grains or particles. Sweeney and Kaplan (1973) provide a succinct account of their probable origin.
Examples: pyrite, uraninite.

6 PARAGENESIS

6.1 INTRODUCTION

The paragenetic sequence of an ore deposit is the order of formation of the minerals. Normally it gives the time succession (i.e. oldest to youngest or earliest to latest) of events which gave rise to and affected that deposit or ore field. It should also describe the conditions under which each phase was formed or was re-equilibrated. In addition to the ore minerals, the paragenetic sequence should note the time at which the gangue minerals started to be deposited and the duration of their deposition. The sequence of events is often represented as a bar-chart diagram or flow diagram (Fig. 6.1). Picot and Johan (1982), and other European authors, use the term paragenesis as a synonym for the mineral assemblage and/or mineral associations. In this book, however, it is used *only* to describe a time succession. Not all ore deposits are capable of being paragenetically interpreted. For example, an ore deposit may have been subjected after its initial deposition to either prograde or retrograde metamorphism. The original depositional features may have either been destroyed or so altered as to mask or mislead an investigator studying the deposit. Some lead–zinc deposits, which are classified as being of a high-temperature origin, may in fact have deposited in a low-temperature regime, but due to subsequent metamorphic effects all the original features of a low-temperature origin may have been destroyed. Arguments as to the true nature of the initial deposition may never be resolved.

In order to deduce a paragenetic sequence, the examination of the *in situ* ore as well as polished samples should ideally be undertaken. Often it is impossible, however, to undertake the former requirement as the locality may be 'lost', abandoned, flooded, etc. However, the aim is as far as possible to:

> Identify the phases present, recognise the textures exhibited in the samples or in the mine or outcrop and report on the features which may help to diagnose the time-related events.

This information, together with that obtained by other geological and mineralogical techniques, can be used to elucidate the geological history of the ore, including the correlation of the different parts of the deposit. It may be possible to obtain supplementary information from studies of trace element variations, stable and radiogenic isotope abundances, fluid inclusion analyses, etc.

As each ore deposit is unique, there is no correct or standard procedure for erecting a paragenetic sequence of mineralisation. It relies on a representative

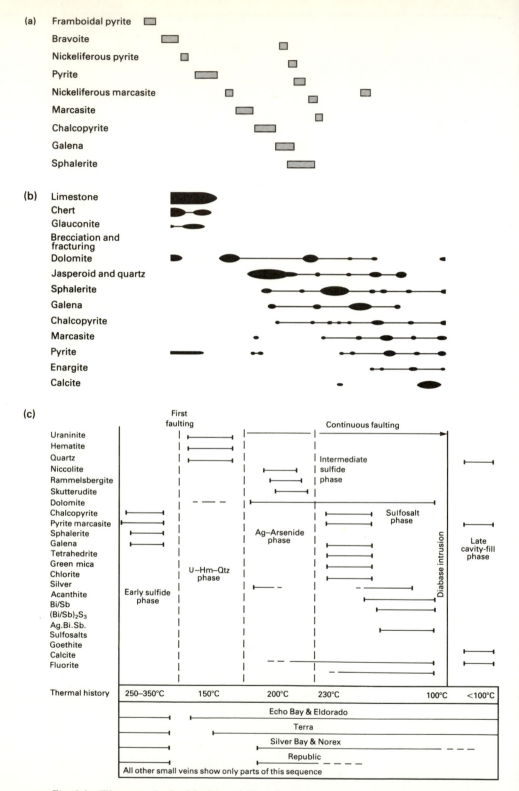

Fig. 6.1 Three methods, (a), (b) and (c), of representing a paragenetic sequence of ore minerals (after D.J. Vaughan and R.A. Ixer 1980; R.D. Hagni and O.R. Grawe 1964; and J.P.N. Badham *et al.* 1972).

suite of samples being collected and this in itself is, possibly, the most difficult part of the study. In that a limited number of samples have to represent the whole deposit, it is paramount that they represent all the various rock types and wallrock lithologies. In some circumstances, orientated samples and larger than normal polished blocks need to be studied. Polished thin sections are often informative in that they are able to illustrate the inter-relationships between the ore minerals, the gangue minerals and the host lithologies. Polished thin sections are also better for revealing the internal structure of certain ore minerals (e.g. sphalerite, tetrahedrite, pyrargyrite, etc.) than are the more 'standard' techniques.

6.2 AIDS IN ERECTING A PARAGENETIC SEQUENCE

6.2.1 CRYSTAL SHAPE

In general, euhedral (idiomorphic) crystals have grown early in an unobstructed medium or an open space (e.g. vein, vug or cavity). Certain minerals are commonly found to crystallise in idiomorphic form which is often related to their specific depositional environment or position in a general paragenetic sequence; such minerals include: fluorite, galena, sphalerite, quartz, cassiterite and covellite. They may be mixed with or overgrown by other minerals or overlain by the same mineral of a later generation. The overall shape may indicate the direction of growth, for euhedral faces and growth zones, if present, *must* have formed in the growth direction. *Stoss-side growth* (see 5.2.1(a)) will, if observed in the sample, substantiate this growth model.

The shape of the individual faces may also help, for *convex* faced crystals have often been interpreted as being formed before crystals with *concave* faces. This feature is often noted in cubes and octahedra. As with all generalities, there are exceptions to the rule. Pyrite and arsenopyrite are a case in point for these two minerals frequently form euhedrally in all environments, e.g. pyrite cubes in slate, whether early or late. However, a mineral may not be a phase of primary growth but may be the result of subsequent dynamic and/or thermal effects to which a mineral suite has been subjected. The metamorphic recrystallisation of a deposit may result in the formation of cubes or pyritohedra porphyroblasts within a pre-existing mineral assemblage. Sometimes a complete geological history of a deposit will have to be ascertained before it can be stated confidently that a particular mineral was formed by a specific geological event.

6.2.2 RELICT (PSEUDOMORPHIC) COMPLEXITIES

A mineral may have formed with euhedral or subhedral crystal outlines at an early stage in a paragenetic sequence, but may subsequently be corroded, replaced or pseudomorphed by other mineral phases and therefore be difficult to position in the sequence of events. In such circumstances, relict structures can sometimes be recognised or inferred (see 5.4.4). An example of this kind would be the presence of crystallographically orientated inclusions in a mineral which does not normally show such features, but which is commonly observed as a pseudomorph of a pre-existing mineral.

6.2.3 MUTUAL GRAIN BOUNDARY PROBLEMS

Mutual grain boundaries between two or more phases often indicate an equal degree of penetration into one another. In such an instance, which is very common, the inter-relationships are very difficult to confirm, particularly if there are no replacement textures or characteristically first-formed crystals present. Normally, it is not possible to deduce a sequence for these minerals and they may indeed have been deposited simultaneously.

6.2.4 COLLOFORM BANDING

Colloform, concentric banding or concentric botryoidal overgrowth structures grow outwards from a nucleus. Each growth period is indicated by an overgrowth band which may be due to a change in the size, shape, orientation or colour at different stages of deposition, or to the deposition of another mineral.

Such features indicate changes, which may be quite small, in the ore fluid or environment of precipitation. Bands may develop outward from the wallrocks, overlie a previous non-colloform depositional phase or occur as discrete spherical growths (pisoliths, oolites, etc.). Roedder (1968) reported that colloform sphalerite was the result of crystallisation of fine fibrous crystals from a fluid and not, as previously assumed, associated with gel formation.

Colloform texture is more frequently encountered in open-space-filling ores, and as such is common in iron and manganese oxides, uranium minerals, arsenides as well as pyrite and sphalerite. The examination of polished thin sections in transmitted light often shows the minute (micrometre scale) colour banding and indicates that single crystals may contain hundreds of growth bands.

6.2.5 GROWTH ZONING

Individual crystals may show growth zones which are seen as colour bands indicating that the environment was changing during formation. Such features occur not only in hydrothermally deposited vein minerals but also in magmatically precipitated minerals such as chromites and magnetites.

6.2.6 CROSS-CUTTING RELATIONSHIPS

This is possibly one of the most important aspects in erecting a paragenetic sequence, and can be one of the easiest to observe. If one feature cuts another, then it must be younger than that which it cuts, *except* if the first feature has been replaced *or* where both features result from remobilisation. The feature may be a vein, veinlet, fault (infilled or not) or a sedimentary structure, all of which frequently occur in syngenetic ores.

6.2.7 TWINNING

Growth twins may develop in some grains and not in others and may differentiate mineral generations. Inversion twinning is more difficult to recognise but, if seen, indicates a sequence of falling temperature and at least partial re-equilibration. Deformational twinning may occur at any stage and can indicate that varying deformational processes were active during the crystallisation history of the deposit. It may indicate an early deformation, if present in one or more primary mineral phases, or if located in minerals of all stages it shows that deformation post-dates mineral deposition.

6.2.8   EXSOLUTION

Exsolution intergrowths provide the best evidence that simultaneous deposition has taken place. On segregation, a granular or allotriomorphic texture, also called 'mutual-boundary texture', occurs. The two minerals have smooth, curved contacts without projections into each other. This texture is not exclusively developed due to exsolution for it can be formed by replacement processes. Zonal textures formed from a gel and rapid alternations in crustified banded textures can also indicate simultaneous deposition.

If cooling has been very slow, it may have allowed the complete unmixing and segregation of the precipitated minerals. This may take place if the concentration of the solute was low, when small amounts of the solute mineral can occur as thin films in the grain boundaries of the host as well as at the contact of the host with transparent gangue minerals. Such textures indicate simultaneous deposition, unless evidence to the contrary is available.

6.2.9   REPLACEMENT

As the mineral which is being replaced must have formed before the 'secondary mineral' which is replacing it, the paragenetic sequence is obvious. Replacement textures and weathering processes are often initiated by a surface reaction on crystal boundaries and in fractures and as such will initially form an outer skin or rim to one mineral phase. However, all stages of replacement are possible from the initial and incipient to total (relict/pseudomorphic) replacement. In an advanced form, the replacing mineral *may* possess convex boundaries, while the replaced phase may show concave boundaries. In some cases, remnants of the original phase may be present, showing an 'island texture' consisting of residual fragments of the first mineral in a matrix of the later phase.

The most difficult aspect of using replacement phenomena in erecting a paragenetic sequence is when total replacement has taken place. Recognition of complete replacement may be helped by the presence of a crystal habit or texture that is very characteristic of the replaced rather than the replacing mineral, e.g. 'cubes' of chalcopyrite, covellite or goethite replacing pyrite, or banded goethite replacing marcasite.

6.2.10   FLUORESCENCE

Fluorescence is the emission of visible light due to the exposure of a mineral to ultraviolet (UV) light which may be either 'long wave' (300 to 400 nm) or 'short wave' (less than 300 nm). It may assist in the erection of a paragenetic sequence for rough blocks and *in situ* material as well as for cut and polished samples. A number of minerals, which are difficult to recognise ordinarily, can be distinguished under short-wave UV light. Typical examples are cassiterite, scheelite and some of the common gangue minerals e.g. certain calcites, dolomites and fluorites. Since some minerals which may ordinarily be difficult to recognise give different colour and intensity responses to UV irradiation (particularly of short wavelength), their existence in an aggregate may be revealed by irradiation. Typical examples are calcite, dolomite and fluorite. However, not all the samples of one mineral from the same locality need fluoresce. Fluorescence may also reveal growth zones and other banded features within a mineral that are not normally visible.

6.2.11   ADDITIONAL READING

More detailed accounts of paragenetic sequences and relationships are given, some with extensive photographic illustration, in such works as: Bastin (1950), Oelsner (1966), Maucher and Rehwald (1961 onwards), Uytenbogaardt and Burke (1973), Ramdohr (1980) and Craig and Vaughan (1981).

7 STRUCTURAL ETCHING

7.1 GENERAL COMMENTS

Although etch tests are not used as frequently as they used to be, they can still perform a useful function in ore microscopy. The demise of the diagnostic etch tests is partially due to the development of more definitive methods of analysis and identification, e.g. the use of the electron microprobe and the application of X-ray diffraction and fluorescence techniques. Etching should only be considered as one of the many aids towards the identification of a mineral. Structural (or textural) etching is commonly used in metallurgical laboratories where it may reveal alloy microstructures which would otherwise be invisible.

7.2 THE METHOD

The technique was first introduced by Schneiderhohn in 1920 and described more fully by Schneiderhohn and Ramdohr (1931). It is often the only way in which the textures of some minerals can be examined or even noted. It may indicate zoning, twinning and grain boundaries in material which before treatment appears to be homogeneous. Phases which may be optically indistinguishable are often differentiated by structural etching; thus, for example, monoclinic and hexagonal pyrrhotite are differentiated when treated with an etch of ammonium dichromate and hydrochloric acid solution.

As most types of etching destroy the surface of the polished specimen which took time and care to produce, it is not a technique to be applied indiscriminately. However, certain samples corrode naturally (tarnish in the air) rather than artificially (by etching) so this process can be made use of as well. By leaving the polished sample out of the air-tight cabinet or desiccator, a tarnish may develop either slowly or rapidly. The silver minerals, and especially argentite, are prone to this phenomenon, which in this case is similar to the darkening of a photographic film. Tarnishing may occur even under the most stringent storage conditions and hence lead to the need for constant repolishing, but it can be turned to good use in the identification of a small number of minerals.

7.3 THE REAGENTS

Table 7.1 lists some of the more frequently used chemicals for this technique.

Table 7.1   Reagents used in structural etching

HNO_3	concentrated – 1 part acid to 1 part water
HNO_3	dilute – 2 parts acid to 3 parts water
HNO_3	dilute – 1 part acid to 7 parts water
Aqua regia	3 parts concentrated HCl to 1 part concentrated HNO_3
H_2SO_4	concentrated – 1 part acid to 1 part water
H_2SO_4	dilute – 1 part acid to 4 parts water
HI	concentrated
HBr	concentrated
KCN	20 g in 100 cc of water
CrO_3	50 g in 100 cc of water (50%)
CrO_3 + HCl	1 part CrO_3 (50%) and 1 part concentrated HCl
$KMnO_4$	2.5 g in 100 cc of water
$KMnO_4$ + conc. HNO_3	1 part $KMnO_4$ to 1 part conc. HNO_3
$KMnO_4$ + conc. H_2SO_4	1 part $KMnO_4$ to 1 part conc. H_2SO_4
$KMnO_4$ + KOH	1 part $KMnO_4$ to 1 part KOH
HCl	concentrated – 1 part acid to 1 part water
HCl	dilute – 1 part acid to 3 parts water
HF	concentrated
$AgNO_3$	saturated solution in water
$SnCl_2$ + HCl	1 part saturated solution to 1 part conc. HCl
$FeCl_3$	20 g in 100 cc of water
$FeCl_3$	50 g in 100 cc of water
KOH	saturated solution in water
K_2CrO_4	saturated solution in water diluted to three times its volume
Picric acid	4 g in 100 cc of ethyl alcohol
$(NH_4)_2Cr_2O_7$	400 mg in 25 cc of 15% HCl
NH_4OH	1 part conc. NH_4OH to 1 part 16% H_2O_2
NH_4OH	5 parts conc. NH_4OH to 1 part 3% H_2O_2

Reproduced, with minor modifications, from Schneiderhohn and Ramdohr (1931), Short (1940) and Craig and Vaughan (1981) with the permission of the publishers.

8 Mineral tables

8.1 INTRODUCTION AND EXPLANATION OF THE TABLES

The following information is given:

International mineral name and variations in spelling as well as synonyms and colloquial names.

Formula and alternative or ideal formula are provided and beneath which are noted any solid solutions, major substitutions in the lattice or impurities if any, that may be present.

Crystal system (or systems, if a temperature inversion occurs).

MEGASCOPIC FEATURES

Colour: The colour noted on a fresh surface with or without variations.

Streak: The colour of the powder produced with scratching or rubbing on an unglazed porcelain plate or roughened glass.

Lustre: Not a strong diagnostic feature but one which may assist in the identification. It depends on the surface texture and whether the mineral is liable to alteration (tarnishing or oxidation).

MICROSCOPIC FEATURES

Colour: *In general:* The overall visible colour a monominerallic sample should show when viewed under low magnification and at optimum illumination.
In air: The colour of a heterogeneous ore sample depends on its surrounding minerals. The → means that the sample's tint is that colour if it is next to or surrounded by the other mineral specified (otherwise, see Table 4.1).
In oil: Often a more positive aid, but must be compared to that sample's colour in air relative to 'in oil' immersion.

Note: The colour depends on the degree of reflectance and can vary with the degree of polishing, granularity, admixture with other ores or gangue minerals present. The magnification and the intensity of the light source are important contributory factors to either the correct or incorrect colours being shown.

Bireflectance/pleochroism: A description, in note form, of any observable bireflectance and reflection pleochroism. Normally reported for 'in air' observations, but may be recorded 'in oil'.

Crossed polars anisotropy: A description of the presence, intensity and character of any anisotropism with perfectly crossed polars or with slightly uncrossed polars.

Internal reflections: Indicates whether they may be shown and, if so, the most frequently noted colours.

Textures: A guide, and only that, to some of the more common textures which may be exhibited by the mineral.

Associated minerals: An aid to identification, for an associated mineral may be recognisable and lead to ideas as to the possible presence of other paragenetically related ores.

Other remarks: Useful notes, hints, warnings and advice etc. that may aid the identification.

Density: These values (g/cm^3) are given for comparative purposes and also methods of separating by heavy liquid methods where appropriate.

Reflectance (%) in air: At 546 nm; and at 589 nm
(see Tables 4.5 and 4.6; also note that the commonly employed wavelengths are those quoted, although any wavelength may be used.)
The mean and the range of the reflectances are also included. For details of the reflectances 'in oil', see other texts, e.g. Ramdhor (1980).

Hardness (Mohs'): The value is given for comparative purposes; no value is given if it is not known with some confidence.

Hardness (talmage/polishing): The initial figure quoted is Talmage's hardness value, but for some minerals he did not quote a value. Only with imperfectly polished, old or worn samples will marks be noticed. Some samples, however, are virtually impossible to polish to perfection. The information gives indications as to the relative hardness with respect to other frequently associated minerals. ($>$ = greater than; $<$ = less than and $\sim$ = approximately equal to)

VHN: A mean and a range is given because, although some texts indicate the hardness with respect to a specific load, this can be misleading since the value can and often does vary with orientation (if parallel or at right angles to a cleavage), grain size and choice of indentation site.

Diagnostic features: A note of some of the more common, but by no means the only, features which may help in the identification. Exotic or unusual forms of a mineral may occur and all the possibilities cannot be covered in this book.

Occurrence: A note as to the general environments of occurrences throughout the world (see also Appendix 1). The selection of typical occurrences is, however, very subjective.

Etch tests: Structural etch tests are indicated when appropriate (see Table 7.1).

The data presented in these tables have been derived from the following sources, which should be consulted for further detailed information on these and other minerals: Cameron (1961), Craig and Vaughan (1981), Edwards (1960), Henry (1977), McLeod and Chamberlain (1968), Picot and Johan (1982), Ramdohr (1980), Shouten (1962), Short (1940), Talmage (1925), Tarkian (1974) and Uytenbogaardt and Burke (1973).

ARGENTITE

Ag_2S Monoclinic (>179°C inverts to cubic)
(Silver Glance, **ACANTHITE**)

Colour: Normally dark-grey to lead-black; often black with a sooty-like tarnish. May have a greenish tint.

Streak: Blackish lead-grey.

Lustre: Dull metallic.

Colour:

In air:	*In oil:*
Generally: light-grey, may show a greenish colour if next to white minerals.	Generally: dull-grey, greenish colour, more distinct.
→ Galena: darker, greenish grey tint.	→ Galena: _____
→ Silver: dark greenish grey.	→ Silver: _____

Bireflectance/pleochroism: Very weak. Lower than galena but higher than tetrahedrite–tennantite.
May appear dull even in oil. Better to examine grain boundaries and twin lamellae.

Croosed polars anisotropy: Observable, even if very weak. If polars are just uncrossed, then definite colours are seen, i.e. purplish slate-blue to greenish. If polars fully crossed, appears isotropic.

Internal reflections: Not present.

Textures: Commonly found as granular aggregates or as coatings of fine grains. May be found as loose masses. Often shows twin lamellae due to inversion from the cubic to the monoclinic form; may be used as an imperfect geothermometer.

Associated minerals: Chalcocite, sphalerite, galena, pyrite, silver, Co–Ni arsenides, etc.

Other remarks: Freshly polished samples should be used. May be mistaken for bournonite if this mineral occurs in rounded grains. Scratches are a common feature. Argentite is the high-temperature form of acanthite, hence most natural specimens are acanthite pseudomorphs after argentite.

Density (g/cm³): 7.0–7.2

Reflectance (%) in air:

at 546 nm	*at 589 nm*
30.3–31.3	29.0–29.8
Mean: 34.2; range: 29.0–36.0	

Hardness (Mohs'): 2.0–2.5

Hardness (Talmage/polishing): **A** (less than most other minerals).

VHN: Mean: 42; range: 20–91.

Diagnostic features: It is soft and malleable and therefore difficult to polish. Scratches are always present. Often occurs in individual crystals of an unusually large size. Difficult to recognise only when comparisons cannot be made with other minerals.

Occurrence: Often found in near-surface veins (epithermal), alone or with Ag–sulphosalts and quartz, calcite, barite, etc. May occur with sphalerite, galena, pyrite, gold, etc. Noted in Co–Ni–Ag–V–Bi veins. Widespread in the secondary sulphide zones of enrichment and some oxidation zones. The silver is soluble as the sulphate or carbonate in the overlying oxidised 'iron-cap', and is precipitated as the sulphide by other metals in the secondary sulphide zone.

Typical occurrences are: Broken Hill (NSW, Australia), Mt Isa (Queensland, Australia), Freiberg (E. Germany), Comstock (Nevada, USA), Guanajuato (Mexico), Colquechaca (Bolivia), Jachymov (Czechoslovakia).

Etch tests: HNO_3: light-brown stain.

HCl: tarnishes; aureole will not wash off; a good test.

KCN: stains brown to black.

$FeCl_3$:   stains grey to black.

$HgCl_2$: may show iridescence.

ARSENOPYRITE

FeAsS   Monoclinic
(Mispickel)
(Always contains some Co (may be up to 12 wt%; also Au and Sb))

Colour: Silver-white to steel-grey. It tarnishes very rapidly. May show creamy or pink tint.

Streak: Dark greyish black.

Lustre: Metallic.

Colour:

In air:	*In oil:*
Generally: white with a tint of yellow or yellow-pink.	Generally: white, tint variations far more distinct.
→ Galena: yellower.	→ Galena: bright-yellow.
→ Pyrite: less bright.	→ Pyrite: no change.
→ Löllingite: very similar.	→ Löllingite: brighter.
→ Silver: darker and whiter.	→ Silver: no change.
→ Cobaltite: lighter and whiter.	→ Cobaltite: no change.

Bireflectance/pleochroism: Weak, but noticeable. White with a yellowish tinge. Better noted in oil.

Crossed polars anisotropy: Moderate to strong. Best seen when polars not perfectly crossed; then it may be light-blue, green, reddish brown or yellow. Sb varieties may show blue to pale-green colours.

Internal reflections: Not present.

Textures: Zonal structures which may be revealed by etching the sample are common but may be mistaken for mimetic twins which occur parallel to {010}. Cataclastic textures are also frequently observed. Lamellae, if present, are parallel to {100} and {001}. If early in paragenetic sequence idiomorphic, if late in sequence idioblastic.

Associated minerals: Galena, sphalerite, cassiterite, tetrahedrite, gold, scheelite, bismuthinite, marcasite, niccolite, cubanite, covellite, chalcocite, uraninite.

Other remarks: Frequently shows replacement by galena, chalcopyrite or sphalerite, initially along cataclastic cracks and sometimes as a core replacement.

Density (g/cm³): 6.0–6.2

Reflectance (%) in air:

at 546 nm	*at 589 nm*
50.2–53.7	50.6–54.4
Mean: 54.7; range: 51.7–55.7	

Hardness (Mohs'): 5.5–6.5

Hardness (Talmage/polishing): F (≫ löllingite, magnetite; < cobaltite, pyrite).

VHN: Mean: 1010; range: 890–1283.

Diagnostic features:	It can be mistaken for all minerals in the pyrite/marcasite group. It is similar to löllingite, but löllingite is softer, etches more readily and is whiter (in air). Occurs commonly in rhomb-shaped crystals. Cleavage is very rarely developed. Lamellar twins reasonably common as are zonal textures.
Occurrence:	Occurs in almost all environments. Common in magmatic associations, pegmatatic suites and contact-pneumatolytic high-temperature sulphides. Gold–quartz veins often contain arsenopyrite. It has also been reported in sedimentary gravels and metamorphic terrains. Typical occurrences are: Freiberg and the Harz Mountains (E. Germany and W. Germany), Sala (Sweden), Skutterud (Norway), Roxbury (Conn., USA).
Etch tests:	A good etch cleavage parallel to {101} is sometimes developed. HNO_3: stains, effervesces and tarnishes brown. H_2O_2: develops structures and tarnishes brown. Stain may wash off with HCl.

BISMUTHINITE

Bi_2S_3 Orthorhombic
(Bismuth Glance) (Se can substitute for S)
              (Forms a solid solution series with stibnite)

Colour:	Tin-white with a tint of lead-grey. May darken with time to a pale-yellow as well as tarnishing and turning iridescent.
Streak:	Lead-grey.
Lustre:	Metallic.

Colour:	*In air:*		*In oil:*	
	Generally: pure white but depends on the orientation.		Generally: it has a blue-grey tint.	
	→ Galena:	yellow-white.	→ Galena:	lighter and may be creamy white.
	→ Chalcopyrite:	bluish grey.	→ Chalcopyrite:	_____
	→ Stibnite:	much lighter.	→ Stibnite:	_____

Bireflectance/pleochroism:	Weak to distinct, but weaker than stibnite. Creamy white/grey-white or bluish grey-white.
Crossed polars anisotropy:	Very strong (especially in oil). Grey to olive in air; grey to violet in oil. Straight extinction. Less anisotropic grains may have a brownish polarisation tint. Large grains may have an undulatory extinction.
Internal reflections:	Not present.
Textures:	It is often an early formed mineral and hence is frequently idiomorphic. May occur as thick columnar or radial aggregates or as needles in voids. May be allotriomorphic when replacing other minerals or where it is the product of decomposition. Exsolution textures are rare, zoning is never seen and twins are very rare indeed.

Associated minerals:	Galena, Bi/Pb sulphosalts, antimony, Co/Ni minerals, magnetite, wolframite, cassiterite, gold, scheelite, arsenopyrite, pyrrhotite.
Other remarks:	May form a solid solution with aikinite ($CuPbBiS_3$) and show unlimited mixed-crystal formation with Sb_2S_3.
Density (g/cm^3):	6.5

Reflectance (%) in air:

at 546 nm	at 589 nm
38.5–48.8	38.1–47.9

Mean: 45.9; range: 42.0–50.0

Hardness (Mohs'):	2.0–2.5
Hardness (Talmage/polishing):	**B+** (> chalcopyrite, > galena). Easy to polish if no hard minerals are present.
VHN:	Mean: 127; range: 92–172.
Diagnostic features:	Can be confused with numerous other minerals. The most important characteristics are: (a) the pure white colour; and (b) that it has a higher anisotropism than the other minerals with which it may be confused, i.e. boulangerite.
Occurrence:	Mainly found in high-temperature (i.e. contact metamorphic) deposits, e.g. with tin ores, with wolframite or in gold veins. Common in hydrothermal deposits of the Ni–Co–Ag vein type as well as Ag–Sn veins. Has been reported from placer deposits. Typical occurrences are: Bolivia, Boliden (Sweden), Cornwall (UK), Beaver Co. (Utah, USA).
Etch tests:	HNO_3: blackens but rubs off easily. HCl: (if conc.) iridescent and effervescent. $HgCl_2$: brown effervescence, rubs off to give a light-brown colour.

BORNITE

Cu_5FeS_4 probably Tetragonal
(Peacock Ore) (Cu_3FeS_3 to Cu_9FeS_6)
(Purple Copper Ore etc.) Cubic above 228°C.

Colour: Rose-brown or pink-brown (similar to pyrrhotite when fresh surface examined). Very quickly tarnishes from copper colour to violet and blue (hence called 'peacock ore').

Streak: Pale-greyish black.

Lustre: Metallic.

Colour:

In air:	In oil:
Generally: pink-brown, tints towards violet and blue.	Generally: colour changes to orange.
→ Enargite: more coloured and variable.	→ Enargite: _____

Bireflectance/pleochroism: Rather low. Lower than tetrahedrite–tennantite. Higher than sphalerite. Examine grain boundaries and twin lamellae.

Crossed polars anisotropy: Weak, **but** variable anisotropy. Can be very distinct (especially in oil and at low magnifications). Under uncrossed polars is brown to grey to pink-brown in colour. Very fine-grained aggregates may appear isotropic.

Internal reflections: Not present.

Textures: Granular aggregates of rounded grains are common, as are recrystallisation phenomena. Twinning {111}, if present; sometimes polysynthetic. Zoning not present. Occasionally deformation effects are visible. Exsolution seen with chalcopyrite lamellae as 'spindles', or bornite exsolving from chalcocite.

Associated minerals: Chalcopyrite, chalcocite (often in fractures), covellite, tetrahedrite–tennantite, sphalerite, cassiterite, tellurides.

Other remarks: Exsolution and replacement textures may have the same appearance. It often has inclusions (of replaced minerals), especially of enargite, cubanite, cobaltite, magnetite and sphalerite.

Density (g/cm³):   5.0–5.1

Reflectance (%) in air:

at 546 nm	at 589 nm
18.5–21.3	21.5–24.5
Mean: 22.7; range: 21.0–23.9	

Hardness (Mohs'):   3.0

Hardness (Talmage/polishing):   **D+** (> galena, chalcocite, < chalcopyrite).

VHN: Mean: 103; range: 84–124.

Diagnostic features: Its colour (reddish brown) together with low anisotropy is so characteristic as to be virtually unique. Bornite is **not** anisotropic. Pyrrhotite, with which it may be confused, is sometimes harder and brighter in colour. Polysynthetic twins and purplish tarnish are diagnostic.

Occurrence: Widespread geographically and genetically.
1. Magmatic deposits with pyrrhotite.
2. Sedimentary ores not as common but has been noted in 'red-bed' environments.
3. Metamorphic ores resulting from dynamic metamorphism.
Typical occurrences are: Cornwall (UK), Ross Island (Eire), Butte (Montana, USA) and various localities in Bolivia, Peru and Mexico.

Etch tests: HNO_3: brownish yellow; effervescence.
KCN: stains brown.
$FeCl_3$: orange tint.
May be etched by standing it in a humid atmosphere.

BOULANGERITE

$Pb_5Sb_4S_{11}$ (approx. formula)　　　　Monoclinic
(or $5PbS.2Sb_2S_3$)
(May form a solid solution series with falkmanite $Pb_3Sb_2S_6$)

Colour: Dull lead-grey with a bluish tint. Often covered with yellow spots due to alteration.

Streak: Brownish.

Lustre: Metallic.

Colour:

In air:	In oil:
Generally: galena-like, may have light blue-green tint.	Generally: _____
→Galena: very similar or dark greenish grey.	→ Galena: noticeably darker with a distinct grey-green colour.
→Stibnite: similar or lighter.	→ Stibnite: lighter, changes colour more in oil immersion.

Bireflectance/pleochroism: Weak, but noticeable in unorientated aggregates. In air it is lighter parallel to the fibres while perpendicular to them it is darker with a blue-green tint. In oil: more distinct.

Crossed polars anisotropy: Distinct and even more so in oil with an intense illumination. Tan, brown or bluish grey colours.

Internal reflections: Rare, but if noted are indistinct and red in colour.

Textures: Always fibrous. Twinning and zoning have not been observed. Characteristic are the irregular intergrowth boundaries (especially with galena). If weathered, galena may form in boulangerite aggregates.

Associated minerals: Galena, stibnite, pyrite, sphalerite and jamesonite (a rare mineral).

Other remarks: It may be confused with jamesonite, but does not react with KOH. Falkmanite (a variety of boulangerite) is characterised by compact crystals and indentations penetrated with galena.

Density (g/cm^3):   5.6–6.5 (highly variable)

Reflectance (%) in air:

	at 550 nm	at 579 nm
$R_p//c$	37.5	36.5
$R_g\perp c$	41.5	40.4

Mean: 40.2; range: 37.0–44.1

Hardness (Mohs'):   2.5–3.0

Hardness (Talmage/polishing):   **B+** (galena, < bournonite).

VHN: Mean: 148; range: 116–217.

Diagnostic features: May be confused with jamesonite and falkmanite. Galena is isotropic while bournonite forms rounded grains and has twin lamellae as well as a lower reflectance. Stibnite is, possibly, one of the few minerals with which it could be easily confused, but stibnite is strongly anisotropic and capable of being etched by KOH. If jamesonite and falkmanite are present, it may be difficult to distinguish these minerals (Ramdohr 1980: 772). Identification should be confirmed on a number of grains.

Occurrence: Widespread geographically. Only observed in hydrothermal deposits, but at all temperatures and pressures. Common in late-stage depositional environments. Typical occurrences are: Sullivan Mine (BC, Canada), Rosebery (Tasmania), Boliden (Sweden).

Etch tests: Distinguished from jamesonite which does react with KOH.
HNO_3:   blackens and effervesces (may tarnish some specimens).

CASSITERITE

SnO_2 Tetragonal
(Tin Stone) (Fe^{3+}, Nb and Ta may substitute for Sn)

Colour: Light-brown; occasionally grey, light-yellow or colourless.

Streak: White, greyish or brown.

Lustre: Greasy to adamantine.

Colour:

In air:		*In oil:*	
Generally: light-grey to brownish grey.		Generally: pale and very light-grey.	
→ Sphalerite:	darker-grey.	→ Sphalerite:	considerably darker.
→ Wolframite:	darker and as brownish.	→ Wolframite:	considerably darker.
→ Stannite:	greyish brown.	→ Stannite:	_____
→ Ilmenite:	_____	→ Ilmenite:	_____
→ Rutile:	brownish.	→ Rutile:	_____
→ Magnetite:	grey.	→ Magnetite:	_____

Bireflectance/pleochroism: Very weak in air and normally only seen on grain and twin boundaries. More prominent in oil but still weak. Pleochroism weak, even if it is strong in transmitted light.

Crossed polars anisotropy: Very distinct, although the colours are not very characteristic and, if present, are masked by internal reflections.

Internal reflections: Very common indeed. Colours are white, yellow and yellow-brown. May show red colours (due to dissolved Fe_2O_3).

Textures: Twinning on {101} common in most cases, but absent in 'needle tin' and colloform masses. Zoning exhibited by the internal reflections and exsolved bodies. Exsolution textures frequently seen. Columbite may exsolve from cassiterite. Cataclastic textures often noted.

Associated minerals: Quartz, chalcopyrite, stannite, hematite, arsenopyrite, wolframite, bismuthinite, molybdenite, magnetite, galena, sphalerite, bornite.

Other remarks: Good relief.

Density (g/cm³):   6.98–7.02

Reflectance (%) in air:

at 546 nm	at 589 nm
11.5–12.4	11.3–12.2
Mean: 12.0; range: 11.0–12.9	

Hardness (Mohs'):   6.0–7.0

Hardness (Talmage/polishing):   **G+** (> quartz, < pyrite).

VHN: Mean: 1201; range: 992–1491.

Diagnostic features: Hardness, poor polish and reflectance as well as the numerous internal reflections are all typical of cassiterite. Pseudo-replacement textures may occur due to cataclastic effects. Cassiterite replaces numerous other minerals (quartz, felspars, topaz, tourmaline, stannite) and pseudomorphs after pyrite, arsenopyrite, sphalerite and bismuthinite occur.

Occurrence: Typical occurrences are: granites in Nigeria; pegmatites in S.W. Africa and Bolivia; veins in Europe and S. America; contact metamorphic in Europe, Malaysia and Alaska; medium-temperature sulphide veins in E. Australia and Bolivia. weathering zones in Bolivia.

Etch tests: No reaction with virtually all reagents, and this helps to differentiate it from rutile, wolframite and uraninite.

CHALCOCITE

Cu_2S Orthorhombic $\rightleftharpoons$ Hexagonal at 103°C
(Digenite, Djurleite) (one of a number of semi-independent minerals and solid solutions).
(Copper Glance)

Colour: White to grey-white with or without a blue tint. May tarnish to a bluish or greenish tint.

Streak: Lead-grey to blackish.

Lustre: Metallic.

Colour

In air:		*In oil:*	
Generally: white (bluish).		Generally: bluish white.	
$\rightarrow$ Galena:	bluish white.	$\rightarrow$ Galena:	without doubt blue.
$\rightarrow$ Pyrite:	bluish grey.	$\rightarrow$ Pyrite:	_____
$\rightarrow$ Covellite:	white (no pink tint).	$\rightarrow$ Covellite:	_____
$\rightarrow$ Bornite:	bluish white.	$\rightarrow$ Bornite:	_____
$\rightarrow$ Copper:	bluish grey.	$\rightarrow$ Copper:	_____
$\rightarrow$ Tetrahedrite:	blue.	$\rightarrow$ Tetrahedrite:	_____

Bireflectance/pleochroism: Very weak indeed, may not be discernible.

Crossed polars anisotropy: Weak to distinct. Colours which may be seen vary from emerald green to pink (intense illumination required). May appear isotropic.

Internal reflections: Not present.

Textures: Complex exsolution intergrowths (use high magnification). Replacement textures are common and complex; most frequently exhibits a lamellar texture which is enhanced after etching with HNO_3.

Associated minerals: Covellite, pyrite, chalcopyrite, enargite, bornite, stannite, sphalerite, galena, bismuthinite, cuprite, copper, goethite.

Other remarks: A **very complex** mineral group – for additional reading, see Ramdohr (1980: 441–67).

Density (g/cm³): 5.5–5.8

Reflectance (%) in air:

at 546 nm	*at 589 nm*
33.1–33.4	31.5–31.8
Mean: 29.0; range: 18.0–33.5	

Hardness (Mohs'):   2.5–3.0

Hardness (Talmage/polishing):   **B** ($\gg$ argentite, galena, $<$ bornite, tetrahedrite).

VHN: Mean: 74; range: 58–98.

Diagnostic features: Few, unless comparative minerals are present. Note blue colour and tarnish as well as polishing scratches. It may show co-relationships with covellite. The hexagonal form (under crossed polars) shows inversion twins. The unusual cubic form (Cu_9S_5) is described by Ramdohr (1980: 442–4).

Occurrence:
1. From hypogene solutions at $> 103\,°C$
2. From supergene solutions at $< 103\,°C$
3. From supergene solutions as a cement on other sulphides, etc.

Etch tests: The etch cleavage {001} is different from the natural cleavage {110}.

HNO_3:	effervesces; produces blue tarnish and parallel etch.
KCN:	blackens and structurally etches.
$FeCl_3$:	dissolves, discolours and stains blue.
$HgCl_2$:	weak etch sometimes noted.

CHALCOPYRITE

$CuFeS_2$ Tetragonal
(Copper Pyrites) (At high temperatures FeS can enter structure; may also accept ZnS, NiS, CoS and SnS)

Colour: Yellow; may appear to have a greenish to greenish yellow tint when directly compared with pyrite.

Streak: Greenish black.

Lustre: Metallic.

Colour

In air:		In oil:	
Generally: yellow.		Generally: very pale yellow.	
→ Galena:	more yellow.	→ Galena:	darker yellow.
→ Pyrite:	deep yellow.	→ Pyrite:	deep yellow and darker.
→ Tetrahedrite:	light yellow.	→ Tetrahedrite:	very bright yellow.
→ Stannite:	light yellow.	→ Stannite:	very bright yellow.
→ Pyrrhotite:	yellow.	→ Pyrrhotite:	pale yellow.
→ Gold/silver:	duller and greenish yellow.	→ Gold/silver:	dull, green to olive green.
→ Sphalerite/ magnetite:	bright yellow white.	→ Sphalerite/ magnetite:	bright yellow-white.

Bireflectance/pleochroism: Pronounced in pleochroic samples which are invariably of a high-temperature origin, but otherwise weak to extremely weak.

Crossed polars anisotropy: Weak to extremely weak but more noticeable in oil (use intense illumination) when grey-blue and greenish yellow colours should be observed.

Internal reflections: Not present.

Textures: Invariably polysynthetic twins are seen. Sphalerite and stannite (sometimes) may exsolve out of chalcopyrite as four-pointed stars and suggest a depositional environment >300 °C.

Associated minerals: A wide range of minerals, e.g. pyrrhotite, galena, sphalerite, pyrite, cubanite, stannite, arsenopyrite.

Other remarks: Only gold has such a yellow colour.

Density (g/cm³):   4.1–4.3

Reflectivity (%) in air:

at 546 nm	at 589 nm
42.5–45.8	44.5–47.6
Mean: 43.2; range: 42.5–44.0	

Hardness (Mohs'):   3.5–4.0

Hardness (Talmage/polishing):   **C** (> galena, < sphalerite, pentlandite and pyrrhotite).

VHN: Mean: 203; range: 174–245.

Diagnostic features:	Easily scratched, while pyrite is not so easily scratched. The colour is characteristic, but only on freshly polished surfaces. Small grains may be confused with gold. Millerite is cream-yellow and lighter in colour. Cubanite is isotropic and has a brown tint. Pyrite is not as deeply coloured.
Occurrence:	Very widespread both geographically and genetically. In magmatic, pegmatitic, pneumatolytic, hydrothermal, sedimentary and metamorphic deposits and in meteorites. Typical occurrences are: Sudbury (Ontario, Canada), Concepcion del Oro (Mexico), Aberfoyle (Tasmania), Kupferschiefer (Poland) and Rammelsberg (W. Germany).

Etch tests:

HNO_3:	darkening of the yellow colour and tarnishing although both or either of these effects may not occur.
$KMnO_4 + H_2SO_4$:	rapidly etches and the coating produced may be difficult to remove.
$KMnO_4 + KOH$:	produces a thick brown coating which may be removed by conc. HCl.

CHROMITE

'$FeCr_2O_4$'   Cubic
actually – $(Fe,Mg)(Cr,Al,Fe)_2O_4$ ($\pm$ Mn, Zn, Ti etc.)

Colour:	Iron-black to brownish black but may be yellowish red in thin section.
Streak:	Brown.
Lustre:	Submetallic to metallic.

Colour

In air:	*In oil:*
Generally: grey-white to light-brown.	Generally: pale-brown.
→ Magnetite: darker.	→ Magnetite: very dark.
→ Sphalerite: darker.	→ Sphalerite: slightly darker.
→ Ilmenite: darker and less red-brown.	→ Ilmenite: very dark and will appear to be brownish red in comparison to chromite.

Bireflectance/pleochroism:	Not present.
Crossed polars anisotropy:	Isotropic but weak anisotropism may appear and especially so in zoned crystals, Zn-rich chromites and tectonically-deformed chromite.
Internal reflections:	Often visible in oil and, when so, are reddish brown in colour (common in Mg and Al varieties). Pure Fe chromites are opaque.
Textures:	Euhedral grains are very common. May form myrmekitic intergrowths with gangue minerals, may show lamellar exsolutions of ilmenite and/or rutile.
Associated minerals:	Pentlandite, pyrrhotite, magnetite, millerite, ilmenite and the 'platinoids'.
Other remarks:	May be feebly magnetic.

Density (g/cm³): 4.32–4.57

Reflectivity (%) in air:

at 546 nm	at 589 nm
11.5–13.3	11.5–13.0
Mean: 12.7; range: 12.0–14.0	

Hardness (Mohs'): 5.5

Hardness (Talmage/polishing): **G**+ (> magnetite, < hematite).

VHN: Mean: 1305; range: 1036–1600.

Diagnostic features: Conchoidal fractures are ubiquitous. Its low reflectance, hardness, isotropism and internal reflections are all good indicators of chromite. Common in ultramafic rocks. It may be confused with magnetite, but the latter is more reflective and has no internal reflections.

Occurrence: One of the earliest minerals to crystallise in a cooling magma. Associated with peridotites and serpentines.
Typical occurrences are: Greece, Turkey, Philippines and S. Africa (Bushveld Complex).

Etch tests: Negative to all reagents.

CINNABAR

HgS

Colour: Cochineal-red to brown-red.

Streak: Scarlet to 'cinnabar-red'.

Lustre: Strongly adamantine.

Colour

In air:	*In oil:*
Generally: white with a blue-grey tint.	Generally: bluish grey.
→ Galena: darker and bluish.	→ Galena: _____

Bireflectance/pleochroism: Distinct pleochroism which varies from yellowish grey-white to brownish grey-white. It is best observed in oil for definitive colours.

Crossed polars anisotropy: Strong, but the tints are masked by the abundant and vivid internal reflections. Light green colours may be seen.

Internal reflections: Numerous and vivid red colours are the most frequently seen although blood-red to 'cinnabar-red' may be shown.

Textures: Generally xenomorphic, but may be euhedral. As it is very soft, it is often deformed and as a result forms sinuous and polysynthetic pressure twins.

Associated minerals: Stibnite, pyrite, opal, marcasite, barite, dolomite, chalcedony.

Other remarks: None.

Density (g/cm³): 8.1

Reflectance (%) in air:

at 546 nm	*at 589 nm*
24.6–29.6	23.8–28.2
Mean: 25.7; range: 24.0–29.0	

Hardness (Mohs'): 2.0–2.5

Hardness (Talmage/polishing): **B±** (> antimony, ≪ cuprite, < galena, pyrite).

VHN: Mean: 77; range: 61–99.

Diagnostic features: Moderate reflectance and abundant internal reflections as well as green polarisation colours are helpful guides. It is often associated with stibnite.
cf. cuprite which is harder.
cf. proustite which is never twinned and has no green polarisation colours.

Occurrence: Restricted.
Formed in late-stage hydrothermal precipitates at 100 °C (i.e. active hot springs). Typical occurrences are: Almaden (Spain), Bakmut (USSR), New Almaden (California, USA), Brewster (Texas, USA).

Etch tests: Negative to all reagents.
Aqua regia may slowly effervesce and stain with iridescent colours.

COBALTITE

(Co,Fe)AsS	Orthorhombic
CoAsS	(Pseudocubic)

(Minor Fe probably due to pyrite/pyrrhotite inclusions)

Colour: White with a distinct pinkish, violet or brown tint which is more noticeable if next to arsenopyrite.

Streak: Greyish white.

Lustre: Metallic.

Colour	*In air:*		*In oil:*	
	Generally: white to cream or even pink.		Generally: no change in oil.	
	→ Pyrite:	whiter.	→ Pyrite:	whiter.
	→ Arsenopyrite:	a pink tint.	→ Arsenopyrite:	a pink tint.
	→ Chalcopyrite:	distinctly pink and lighter.	→ Chalcopyrite:	_____

Bireflectance/pleochroism: Weak with a white to pink tint which is especially seen along twin boundaries.

Crossed polars anisotropy: Weak to distinct (especially in oil) when blue-grey and brown shades can be seen.

Internal reflections: Not present.

Textures: Idiomorphic crystals which occur as inclusions are common. Often the crystals illustrate complex twins and lamellar intergrowths parallel to the crystal faces. (May look like a chess-board.)

Associated minerals:	Chalcopyrite, cubanite, pyrrhotite, sphalerite , arsenopyrite, molybdenite, silver.
Other remarks:	The colour appears to change with time, the pink colour intensifying after being cut and polished. Inversion to cubic, and hence isotropic, form is possible.
Density (g/cm³):	6.3

Reflectance (%) in air:

at 546 nm	*at 589 nm*
49.5–50.5	50.3–51.9
Mean: 53.2; range: 52.0–54.7	

Hardness (Mohs'):	5.5–6.0
Hardness (Talmage/polishing):	**G** (≫ löllingite, > arsenopyrite, < pyrite).
VHN:	Mean: 1179; range: 948–1367.
Diagnostic features:	Pink-white colour; hardness and idiomorphic texture are characteristic as are the weak anisotropy and the possible presence of complex twins. It may crystallise in cubes and pyritohedra.
Occurrence:	Widespread in minor amounts but rare in larger quantities. High-temperature occurrences (i.e. the cubic form) may be due to subsequent metamorphism (i.e. Fahlbands of Norway and Sweden). More frequently noted from medium-temperature deposits of the Co–Ni–Ag–U or Pb–Cu–Hg–Se type. Typical occurrences are: Cobalt (Ontario, Canada), Boliden (Sweden), Cobar and Broken Hill (NSW, Australia), Cornwall (UK).
Etch tests:	HNO_3: some specimens stain, tarnish and are iridescent.

COPPER

Cu   Cubic
(May contain Ag, As, Bi, Hg, etc.)

Colour:	Reddish in freshly cut sections and may have a pink tint but turns to 'copper-red' on tarnishing.
Streak:	Metallic and shining.
Lustre:	Metallic.

Colour	*In air:*		*In oil:*	
	Generally: rose-white.		Generally: no change in oil.	
	→ Silver:	darker and reddish.	→ Silver:	_____
	→ Chalcocite:	brighter.	→ Chalcocite:	even more marked.
	→ Cuprite:	brighter and bluish green to grey.	→ Cuprite:	_____

Bireflectance/pleochroism:	Not present.
Crossed polars anisotropy:	Isotropic, but note that scratches may give false appearances.
Internal reflections:	Not present.
Textures:	Often occurs as inclusions in cuprite. Supergene copper may be dendritic or have

spear-like forms. Fine-grained crystal aggregates may be concretionary or granular. Rhythmic textures occur with limonite (after etching).

Associated minerals:	Cuprite, goethite, chalcocite, covellite.
Other remarks:	None.
Density (g/cm³):	8.5–9.0
Reflectance (%) in air:	

at 546 nm	at 589 nm
60.6	87.0

Mean: 75.8; range: 71.2–81.2

Hardness (Mohs'):	2.5–3.0
Hardness (Talmage/polishing):	**B+** (> chalcocite, chalcopyrite, < cuprite).
VHN:	Mean: 92; range: 48–143.
Diagnostic features:	Extremely high reflectance (when freshly polished), red to pink tint if left to tarnish. Often associated with cuprite.
Occurrence:	Widespread geographically and geologically. Typical occurrences are: Bisbee (Arizona, USA), Lake Superior region (USA), Tetiuhi (Siberia, USSR), Mt Isa (Queensland, Australia), Corocoro (Bolivia), Skaergaard (Greenland).

Etch tests:		
	HNO_3:	effervesces, and sometimes tarnishes.
	HCl:	tarnishes brown (sometimes).
	KCN:	turns brown.
	$FeCl_3$:	a darkening of the colour.
	KOH:	tarnishes brown and becomes iridescent.
	$HgCl_2$:	quickly stains and becomes iridescent.

COVELLITE

CuS   Hexagonal
(Covelline) (May contain Cu_2S in solution)

Colour:	Blue, ranging from dark-blue (indigo blue) to bluish white. Purple when wet.
Streak:	Lead-grey to black.
Lustre:	Submetallic to resinous.

Colour	In air:			In oil:		
		'o'	'e'		'o'	'e'
	Generally:	deep-blue to violet.	bluish white.	Generally:	purple to violet -red.	blue-grey to pinkish.

→ Chalcocite: lighter and pinker.   → Chalcocite: _____
[e // c, o ⊥ c]

Bireflectance/pleochroism:	Extraordinarily high; this is the most characteristic feature.

Crossed polars anisotropy: Extreme. Note the unusually bright orange tints which vary from fiery orange to reddish brown.

Internal reflections: Not present.

Textures: Aggregates of 'twisted' lamellae occur with deformation; also traces of recrystallisation.

Associated minerals: Chalcocite, bornite, enargite, tetrahedrite–tennantite, chalcopyrite, pyrite, stannite, sphalerite, galena.

Other remarks: None.

Density (g/cm³): 4.67

Reflectance (%) in air:

	at 546 nm	at 589 nm
R_o	7.2	4.5
R_e	24.3	21.7

Mean: 14.2; range: 7.0–24.3

Hardness (Mohs'): 1.5–2.0

Hardness (Talmage/polishing): **B+** ($\gg$ argentite, $<$ galena, chalcopyrite, chalcocite).

VHN: Mean: 82; range: 68–110.

Diagnostic features: The very strong pleochroism (blue colour) and extreme anisotropism (orange tints) should avoid any confusion with all other minerals. Se-rich varieties do not show these characteristics, but these are very rare minerals.

Occurrence: Widespread geographically and geologically. Typically in hydrothermal and contact-metamorphic deposits. In sedimentary deposits, mainly found in the zones of enrichment/oxidation, also present in arid alluvial basins. Typical occurrences are: Montana (USA), Tsumeb (Namibia, S.W. Africa), Bisbee (Arizona, USA), Bar (Yugoslavia).

Etch tests: HNO_3: tarnishes.
KCN: produces a black stain.
Aqua regia: temporary effervescence colours the sample red.

CUBANITE

$CuFe_2S_3$ Orthorhombic
$(CuS.Fe_2 S_3)$

Colour: Creamy grey to light (pale) brown. Brass to bronze-yellow also noted.

Streak: _ _ _ _ _ _ _ _ _

Lustre: Metallic.

Colour **In air:**
Generally: yellow-white to rose-brown.

In oil:
Generally: a decrease in the intensity of the colour.

	In air:		In oil:	
	→ Pyrrhotite:	more yellowish, less pink.	→ Pyrrhotite:	no change.
	→ Chalcopyrite:	pinkish grey.	→ Chalcopyrite:	often distinct with cubanite being more yellow.

Bireflectance/pleochroism: Distinct (in oil). The colour is creamy grey to brownish grey.

Crossed polars anisotropy: Strong, but less than pyrrhotite. Has a pinkish-brown, greyish-blue colour. Basal sections are less anisotropic.

Internal reflections: Not present.

Textures: Often occurs as coarse exsolution lamellae bodies in chalcopyrite (**minimum** temperature of exsolution at 250–300 °C). It may also be found as granular or polygonal aggregates.

Associated minerals: Chalcopyrite, pyrrhotite, pentlandite, magnetite, arsenopyrite, ilmenite.

Other remarks: Very characteristic magnetic properties.
(Parallel to 'b' – very high magnetic susceptibility; parallel to 'a' and 'c' – virtually non-magnetic.)

Density (g/cm^3): 4.0–4.1

Reflectance (%) in air:

at 546 nm	at 589 nm
35.4–40.3	37.6–42.4
Mean: 40.5; range: 39.2–42.5	

Hardness (Mohs'): 3.5

Hardness (Talmage/polishing): **C** (> chalcopyrite, < sphalerite, ≪ pyrrhotite).

VHN: Mean: 218; range: 150–260.

Diagnostic features: Magnetic (but less so than pyrrhotite). Distinct pleochroism (brown tints) and bright anisotropy (blue tints). Lamellar texture in chalcopyrite and also lack of relief with respect to chalcopyrite are characteristic.

Occurrences: Widespread and abundant especially in high-temperature environments, i.e. magmatic deposits, pegmatites, pneumatolytic veins and contact-metamorphic deposits as well as hydrothermal veins.
Typical occurrences are: Sudbury (Ontario, Canada), Tetiuhe (Siberia, USSR), Boliden (Sweden).

Etch tests: HNO_3:   may stain light-brown and tarnish.
$K_2Cr_2O_7 + H_2SO_4$:   etching may be used to differentiate it from chalcopyrite.

CUPRITE

Cu_2O Cubic
(Red Copper Ore) (May contain a number of trace elements)

Colour:	Shades of red, sometimes almost black.
Streak:	Brownish red.
Lustre:	Adamantine or submetallic to earthy.

	In air:	*In oil:*
Colour	Generally: light grey with a blue tint.	Generally: much darker with a distinct blue tint.
	→ Copper: darker, grey-blue.	→ Copper: _____
	→ Chalcocite: darker and greenish.	→ Chalcocite: darker and distinctly green.
	→ Hematite: darker with a green-blue tint.	→ Hematite: deeper greenish blue tint.

Bireflectance/pleochroism:	Visible but indistinct so high magnification and illumination may help to detect this feature.
Crossed polars anisotropy:	Green to blue tints visible if polars are slightly uncrossed **but** may be masked by the numerous internal reflections. Best observed in air **not** in oil. This is an unusual feature as the mineral is cubic.
Internal reflections:	Deep red, **always** visible and very characteristic.
Textures:	If the mineral has grown in open spaces, it may show octahedra, cubes or needles. Aggregates are more common as are replacement fabrics (it replaces chalcocite along a fine network texture).
Associated minerals:	Copper, goethite, chalcocite, covellite.
Other remarks:	None.
Density (g/cm³):	6.14

Reflectance (%) in air:

at 546 nm	at 589 nm
26.6	24.6
Mean: 27.7; range: 27.1–28.5	

Hardness (Mohs'):	3.0–4.0
Hardness (Talmage/polishing):	**D−** (> chalcopyrite, copper, < goethite).
VHN:	Mean: 204; range: 179–249.
Diagnostic features:	Very abundant red internal reflections, blue-grey colour, anisotropy and frequent association with copper. It often takes a poor polish.
Occurrence:	Found in the oxidation zone of most Cu-rich deposits and especially where supergene ores are being weathered. Large deposits only found in carbonate environments. Typical occurrences are: Bisbee (Arizona, USA), Talmessi (Iran), Corocoro (Bolivia), Cornwall (UK).

Etch tests:	HNO$_3$:	effervesces and deposits a coating of metallic copper.
	HCl:	deposits a white coating.
	KCN:	darkens and brings out the parallel etch structure.
	FeCl$_3$:	tarnishes and becomes iridescent.

ENARGITE

$Cu_3(As,Sb)S_4$ Orthorhombic
$(Sb/(As + Sb) < 7\%)$

Colour: Grey-black with light violet tint.

Streak: Greyish black.

Lustre: Metallic.

Colour

In air:		*In oil:*	
Generally: pinkish grey to pinkish brown.		Generally: darker than in air, i.e. violet-grey, brownish grey.	
→ Bornite:	pinkish white.	→ Bornite:	_____
→ Chalcocite:	pinkish brown.	→ Chalcocite:	all slightly more intense.
→ Galena:	greyish brown.	→ Galena:	
→ Tennantite:	darker pink.	→ Tennantite:	

Bireflectance/pleochroism: (best examined in oil):
// a greyish pink with a yellow tint.
// b pinkish grey.
// c greyish violet.

Crossed polars anisotropy: Strong polarisation colours which vary from blue to green to red and orange. With uncrossed polars shows multi-coloured hues.

Internal reflections: May occur and if so, are deep-red in colour.

Textures: Always xenomorphic. Prismatic crystals or rounded grains frequently noted. It often shows zonal textures, especially when etched.

Associated minerals: Tennantite, stannite, molybdenite, pyrite, covellite, chalcopyrite, bornite, sphalerite, galena, chalcocite, arsenopyrite.

Other remarks: None.

Density (g/cm^3): 4.4–4.5

Reflectivity (%) in air:

	at 546 nm	*at 589 nm*
// a	25.15	24.41
// b	25.88	26.07
// c	28.72	28.66

Mean: 26.4; range: 24.7–28.1

Hardness (Mohs'): 3.0

Hardness (Talmage/polishing): **D−** (> galena, bornite, chalcocite, chalcopyrite, > tennantite, < sphalerite).

VHN:	Mean: 192; range: 133–358.
Diagnostic features:	Light pink-brown colour, moderate reflectance and highly coloured anisotropy.
Occurrence:	Usually hypogene, but may be supergene. Mainly found in Fe-poor hydrothermal Cu–As deposits. Typical occurrences are: Butte (Montana, USA), Bor (Yugoslavia), Ohkubo (Hokkaido, Japan), Tri-State District (USA).
Etch tests:	HNO_3: a faint tarnish may occur. KCN: stains black and may bring out parallel structure. $HgCl_2$: sometimes stains.

GALENA

PbS   Cubic

Se (up to 18%) and Te may substitute for S, Ag (<0.1%), Sb and Bi may substitute for Pb

Colour:	White (bright), sometimes with a pink tint. If Te is present, then the colour is purplish. Fine-grained samples are darker than coarse-grained ones.
Streak:	Lead-grey.
Lustre:	Metallic.

Colour	*In air:*	*In oil:*
	Generally: pure white, tints due to adjacent minerals.	Generally: strongly decreased in colour.
	→ Sphalerite: white.	→ Sphalerite:
	→ Boulangerite: pinkish.	→ Boulangerite:
	→ Bismuthinite: pinkish.	→ Bismuthinite: } contrast increased for all in oil.
	→ Stibnite: lighter.	→ Stibnite:
	→ Tennantite: pinkish.	→ Tennantite:

Bireflectance/pleochroism:	Not present.
Crossed polars anisotropy:	Isotropic, but a weak anomalous anisotropism is not uncommon (grey to grey-black).
Internal reflections:	Not present.
Textures:	Granular aggregates and skeletal crystals are common. A perfect cubic cleavage is often noted. Twinning may be due to mechanical deformation. Zonal textures are also common, but often are only seen after etching.
Associated minerals:	Almost all minerals, but particularly sphalerite, pyrite, chalcopyrite, tetrahedrite-tennantite, bournonite, pyrargyrite, argentite, arsenopyrite, magnetite, pyrrhotite.
Other remarks:	Often contains inclusions of other minerals (see above) especially the sulphosalts.
Density (g/cm^3):	7.5–7.6

Reflectance (%) in air:

at 546 nm	*at 589 nm*
42.7–43.6	41.9–43.0
Mean: 43.0; range: 42.4–43.2	

Hardness (Mohs'): >2.5 (ductile).

Hardness (Talmage/polishing): **B** (>> argentite, > proustite, covellite, boulangerite; ~ chalcocite, bismuthinite; < bournonite, bornite).

VHN: Mean: 79; range: 56–116.

Diagnostic features: Triangular polishing pits and rather high reflectance. It is also commonly associated with sphalerite and pyrite and may show secondary alteration along cracks and cleavages in cerussite.

Occurrence: Widespread geographically and geologically, but rare in magmatic and pegmatatic deposits. Very common in low-temperature Pb–Zn–Ag deposits, i.e. Mississippi Valley type deposits.

Typical occurrences are: Tri-State (USA), Sullivan Mine (BC, Canada), Pennines (UK), Broken Hill (NSW, Australia), Mt Isa (Queensland, Australia), Zawar (India).

Etch tests: HNO_3: effervesces and blackens.

HCl: tarnishes and becomes iridescent.

$FeCl_3$: stains iridescent, but if bornite present – no reaction.

GOETHITE

α-$Fe_2O_3.H_2O$ Orthorhombic
(Needle Iron Ore) (FeO.OH or $FeHO_2$) ± Al_2O_3, CaO, BaO, SiO_2 etc.

Colour: Grey varying from dull grey to bright grey, may have a bluish tint. The colour is strikingly dependent on the texture and the orientation.

Streak: Brownish yellow to ochre-yellow.

Lustre: Submetallic, sometimes adamantine, silky, fibrous, dull and earthy.

Colour *In air:*   *In oil:*
Generally: grey.   Generally: decreases markedly.
→ Sphalerite: bluish.   → Sphalerite: deep blue.
→ Hematite: much darker.   → Hematite: much darker.
→ Cerussite: ———   → Cerussite: very bright.

Bireflectance/pleochroism: Weak and best examined in oil, but still masked by the internal reflections.

Crossed polars anisotropy: Distinct with bluish tints which are lighter under uncrossed polars. Colours masked by internal reflections. Some colloform varieties are isotropic.

Internal reflections: Abundant, especially in cryptocrystalline forms. Colours are brownish yellow to reddish browns.

Textures: Botryoidal and mammilliform masses. Spherulitic and zoned types may have been formed from a gel.

Associated minerals: Cuprite, copper, pyrite, chalcopyrite, hematite, pyrrhotite, sphalerite, galena.

Other remarks: Pseudomorphs after pyrite are common. Central areas are frequently replaced by other minerals.

Density (g/cm³): 3.3–4.3

Reflectance (%) in air:

at 546 nm   *at 589 nm*
15.5–17.5   15.0–16.6
Mean: 16.7; range: 16.1–18.5

Hardness (Mohs'): 5.0–5.5 (if coarse grained).

Hardness (Talmage/polishing): E (< magnetite, ilmenite and hematite – for crystalline masses).

VHN: Mean: 659; range: 525–824.

Diagnostic features: Colloform and spheroidal textures are common. The bluish anisotropy and the abundant internal reflections are characteristic. Psilomelane has **no** internal reflections and is **more** reflective.

Occurrence: Widespread geographically and geologically. Occurs in the oxidised zone of nearly every ore deposit.

Etch tests: HCl: may tarnish after standing.
Aqua regia: tarnishes.

GOLD

Au   Cubic
(May contain Ag, Pd, Cu, Bi, Pt, Hg, Rh, etc.)

Colour:	Bright or 'golden' yellow which varies with the content of the admixed elements.
Streak:	Gold-yellow.
Lustre:	Metallic.

Colour	*In air:*	*In oil:*
	Generally: 'golden' yellow.	Generally: same as in air.
	→ Chalcopyrite: paler.	→ Chalcopyrite: no change.
	→ Silver: yellow.	→ Silver: no change.
	→ Platinum: more yellow.	→ Platinum: no change.
	→ 'Sulphides': lighter.	→ 'Sulphides': no change.

Bireflectance/pleochroism:	Not present.
Crossed polars anisotropy:	Isotropic (however, never completely in extinction under crossed polars), giving a greenish tint.
Internal reflections:	Not present.
Textures:	It often occurs as inclusions in other minerals. Growth zones are reasonably common as are globular masses. Euhedral crystals are rare.
Associated minerals:	Arsenopyrite, stibnite, cobaltite, löllingite, millerite, the tellurides.
Other remarks:	Pd-rich gold is creamy white. Ag-rich gold (Electrum) is pale-yellow. Cu-rich gold is pink to reddish.
Density (g/cm³):	15.6–19.3 (19.3 when pure, often lower).

Reflectance (%) in air:

at 546 nm	at 589 nm
71.5	83.4
Mean: 73.5; range: 64.0–74.0	

Hardness (Mohs'):	2.5–3.0
Hardness (Talmage/polishing):	**B** (> galena, < tetrahedrite, sphalerite; ~ chalcopyrite).
VHN:	Mean: 65; range: 41–102.
Diagnostic features:	Golden yellow colour and the extremely high reflectance. Takes a poor polish due to its hardness. Chalcopyrite appears greenish grey next to gold. Pyrite may be mistaken for gold especially when pyrite occurs as small grains. A test to distinguish between the two minerals is to treat with $AgNO_3$ solution which attacks pyrite and chalcopyrite etc. but not gold.
Occurrence:	Widespread geographically and geologically, but invariably in small amounts (disseminated). Located in sedimentary, volcanic and igneous environments. Typical occurrences are: Witwatersrand (S. Africa), Kalgoorlie (W. Australia),

Kolar (India), Berezowsk (Urals, USSR), and numerous deposits in the western USA.

Etch tests: KCN: may etch.
Aqua regia: effervesces, stains brown and the fumes cause tarnishing.

GRAPHITE

C   Hexagonal
(Plumbago)

Colour: Brown-grey or lead-black.

Streak: Black.

Lustre: Metallic, dull or earthy.

Colour *In air:*
Generally: white-grey, sometimes with an orange tint.
→ Sphalerite: much lighter.
→ Molybdenite: much darker.
→ Chalcopyrite: darker and brown.
→ Pyrrhotite: darker and brown.

In oil:
Generally: 'o' – no change. 'e' – much darker; nearly black.
→ Sphalerite: _____
→ Molybdenite: _____
→ Chalcopyrite: darker and brown.
→ Pyrrhotite: darker and brown.

Bireflectance/pleochroism: Very strong and characteristic.
'o' – brownish.
'e' – nearly black.

Crossed polars anisotropy: Very strong with straw-yellow to dark brown or violet-grey colours; best seen in yellow light. Basal sections may appear isotropic.

Internal reflections: Not present.

Textures: Plates, blades or laths and sheaf-like aggregates are common. **No** twins or zonal textures noted.

Associated minerals: Virtually every mineral known. Commonly – arsenopyrite, magnetite, hematite, pyrolusite, pyrrhotite.

Other remarks: None.

Density (g/cm³): 2.09–2.23

Reflectance (%) in air:

	at 550 nm	at 589 nm
R_o:	17.4	18.1
R_e:	6.8	7.0

Mean: 14.0; range: 6.0–17.0

Hardness (Mohs'): 1.0–2.0

Hardness (Talmage/polishing): **A** (> chalcopyrite, molybdenite).

VHN: Mean: 9.0; range: 4.0–12.0.

Diagnostic features:	Extreme pleochroism and anisotropism as well as low reflectance. The perfect cleavage is often seen. Molybdenite is more reflective and has a pinkish white colour.
Occurrence:	Widespread in contact and regionally metamorphosed sediments, otherwise it is rare. It is virtually absent in hydrothermal veins, but has been found in high-temperature pegmatitic and pneumatolytic deposits. Typical occurrences are: Huelva (Spain), Lac de la Blanche (Quebec, Canada), Aliber (Lake Baikal, USSR), Franklin (New Jersey, USA).
Etch tests:	Negative to all reagents.

HEMATITE

$\alpha\text{-Fe}_2\text{O}_3$ Trigonal

(Specularite) (May contain $FeTiO_3$ or $MgTiO_3$ in solid solution. Up to 17.64 wt % Mn reported)

Colour:	Dark steel-grey or iron-black.
Streak:	Cherry-red or reddish brown.
Lustre:	Metallic (splendent) but may be dull.

Colour	*In air:* Generally: grey-white with a bluish tint.	*In oil:* Generally: markedly lower and now bluish grey.	
	→ Ilmenite:	white.	→ Ilmenite: _____
	→ Magnetite:	white.	→ Magnetite: no change.
	→ Goethite:	brighter.	→ Goethite: _____
	→ Pyrite:	bluish white.	→ Pyrite: pronounced blue colour.
	→ Chalcocite:	brownish.	→ Chalcocite: more pronounced colours.
	→ Cuprite:	white.	→ Cuprite: white **not** bluish.

Bireflectance/pleochroism:	Weak.
Crossed polars anisotropy:	Very distinct, especially on twin boundaries when grey-blue and grey-yellow colours are seen. Sharp extinction.
Internal reflections:	Deep red and often better seen in poorly polished specimens.
Textures:	Lamellar and twin lamellar textures common for it often occurs as exsolution lamellae in ilmenite. It may replace magnetite along cleavage planes to give martite.
Associated minerals:	Magnetite, ilmenite, rutile, chromite, goethite, pyrite, chalcopyrite, bornite, cassiterite, sphalerite.
Other remarks:	Most important in the banded iron formations where it results from secondary processes.
Density (g/cm³):	4.9–5.3

Reflectance (%) in air:

	at 546 nm	at 589 nm
R_o:	31.0(30.2)	29.6(29.1)
R_e:	24.6(26.1)	23.7(25.1)

Mean: 27.3; range: 24.0–30.6

Hardness (Mohs'):   5.5–6.5

Hardness (Talmage/polishing):   **G** ($\gg$ goethite, magnetite; $>$ ilmenite, $>$ rutile, cassiterite; $<$ pyrite).

VHN: Mean: 948; range: 739–1097.

Diagnostic features: Moderate reflectance and distinct anisotropy are the main features. It is also difficult to polish and should show red internal reflections.

Occurrence: Widespread geographically and geologically. The upper limit of formation is 600–700 °C. Occurs throughout all environments suitable for ore deposition. Typical occurrences are: Kiruna (Sweden), Elba (Italy), Hammersley Range (Australia), Mesabi and Vermilion Ranges (USA), Olonetz (USSR).

Etch tests: Negative to all reagents.
Cold, conc. HF may bring out structure if left in contact for more than two minutes.

ILMENITE

$FeTiO_3$ Trigonal
(Often contains Fe^{3+}, Mn^{2+} and Mg)

Colour: Iron-black.

Streak: Black to brownish red.

Lustre: Submetallic.

Colour

In air:	In oil:
Generally: light to dark brown or dull grey with brown tint.	Generally: lower but still brown and varies on orientation.
→ Magnetite: darker.	→ Magnetite: darker, especially for 'e'.
→ Hematite: much darker.	→ Hematite: much darker.
→ Chromite: lighter and reddish brown.	→ Chromite: difference is pronounced.
→ Sphalerite: brighter, brown.	→ Sphalerite: brighter tint.

Bireflectance/pleochroism: Distinct.
'o' – light pinkish brown.
'e' – very dark brown.

Crossed polars Anisotropy: Strong. The colours vary between light greenish grey to brownish grey.

Internal reflections: Very rare but, if present, are dark brown in colour (common in Mg-rich varieties).

Textures: Typical twin lamellae. Fe-rich varieties have hematite exsolution bodies. It may also occur as exsolution lamellae in magnetite. Granular aggregates are common.

Associated minerals: Hematite, magnetite, rutile, pyrite, pyrrhotite, chromite, pentlandite.

Other remarks: It may be used as an 'imperfect' geothermometer.

Density (g/cm^3): 4.7 ($FeTiO_3$)
4.54 ($MnTiO_3$)
4.05 ($MgTiO_3$)

Reflectance (%) in air:

	at 546 nm	at 589 nm
R_o:	20.1	20.2
R_e:	15.8	16.4

Mean: 19.2; range: 17.0–21.1

Hardness (Mohs'): 5.0–6.0

Hardness (Talmage/polishing): **G** (> magnetite, < hematite).

VHN: Mean: 629; range: 519–739.

Diagnostic features: It has a low reflectance, a brown-grey colour (browner than magnetite) and strong anisotropy. Polysynthetic twins are common. Harder than magnetite, softer than hematite. Rutile has **no** internal reflections.

Occurrence: Ubiquitous and especially so in magmatic (intrusive and extrusive), pegmatitic and other vein deposits. It is rare in hydrothermal deposits.

Common in meteorites.
Typical occurrences are: Bushveld (S. Africa), Lang Ping Hsien (Jehol, China), Sudbury (Ontario, Canada), placer deposits.

Etch tests: Negative to all reagents.

LÖLLINGITE

$FeAs_2$ Orthorhombic
(Co and Ni may replace Fe; S and Sb may replace As)

Colour: Silver-grey to steel-grey.

Streak: Greyish black.

Lustre: Metallic.

Colour:

In air:	*In oil:*
Generally: white, sometimes with yellow tint.	Generally: no change.
→ Arsenopyrite: not as yellow.	→ Arsenopyrite: slightly more yellow.

Bireflectance/pleochroism: Weak, but distinct in oil when a white to bluish yellow-white colour is seen.

Crossed polars anisotropy: Very strong. Bright orange-yellow to reddish brown, pale-brown, blue, slatey blue or green colours may be observed.

Internal reflections: Not present.

Textures: Often occurs as residual inclusions in arsenopyrite. If independent, then six-pointed star-shaped twins are common. Euhedral crystals often show well-developed zoning. A cataclastic texture may result if it is replaced.

Associated minerals: Arsenopyrite, sphalerite, galena, cassiterite, stannite, wolframite, Co–Ni arsenides.

Other remarks: None.

Density (g/cm³):   7.40–7.58

Reflectance (%) in air:

at 546 nm	at 589 nm
52.4–54.1	51.2–55.2
Mean: 53.8; range: 53.0–54.7	

Hardness (Mohs'):   5.0–5.5

Hardness (Talmage/polishing):   F− (≫ chalcopyrite, > pyrrhotite, < magnetite, ≪ arsenopyrite and pyrite).

VHN: Mean: 742; range: 421–963.

Diagnostic features: Its reflectance and strong anisotropy are useful features. It may occur in euhedral form. Marcasite has more highly coloured tints.

Occurrence:	It is present in small quantities in most hydrothermal and contact-pneumatolytic deposits. Typical occurrences are: Boliden and Falun (Sweden), Cornwall (UK), Broken Hill (NSW, Australia), Cobalt (Ontario, Canada), Jachimov (Czechoslovakia), Kowary (Poland), Lölling (Austria).
Etch tests:	HNO$_3$: slowly effervesces and tarnishes iridescent.

MAGNETITE

Fe$_3$O$_4$ Cubic
(Solid solution with Fe$_2$TiO$_4$)
(Fe replaced by Mg, Mn, Zn, Ti, Al, V and Cr)

Colour:	Iron-black.
Streak:	Black.
Lustre:	Metallic (splendent) to submetallic and dull.

Colour	*In air:*		*In oil:*	
	Generally: grey with brown tint.		Generally: markedly decreased.	
	→ Hematite:	darker, no blue-white tint.	→ Hematite:	much darker.
	→ Ilmenite:	brighter, less pink.	→ Ilmenite:	brighter.
	→ Sphalerite:	brighter and browner.	→ Sphalerite:	brighter.

Bireflectance/pleochroism:	Not present.
Crossed polars anisotropy:	Isotropic but with an anomalous anisotropism due to internal strains possibly due to deformations resulting from the formation of zones or twins.
Internal reflections:	Not present, except in samples with a high Mn content.
Textures:	Idiomorphic development is a common feature. Lamellar exsolution of ilmenite. May pseudomorph hematite or be replaced by hematite along cleavage planes (martite). Zoning often noted in pyrometasomatic magnetite.
Associated minerals:	Hematite, ilmenite, sphalerite, chalcopyrite, galena, pyrrhotite, arsenopyrite, cubanite.
Other remarks:	Alters to hematite and goethite.
Density (g/cm^3):	5.2 (variable).

Reflectance (%) in air:

at 546 nm	*at 589 nm*
16.4–20.1	16.9–20.3
Mean: 20.7; range: 20.0–21.1	

Hardness (Mohs'):	5.5–6.5
Hardness (Talmage/polishing):	F (≫ pyrrhotite, < ilmenite, ≪ hematite).
VHN:	Mean: 581; range: 480–734.

Diagnostic features: Magnetic. The absolute and relative hardness. The colour **can vary** considerably. Sphalerite, with which it may be confused, is often rich in internal reflections. Ilmenite is anisotropic in most specimens and is capable of showing numerous special fabrics, see Ramdohr (1981: 911–40).

Occurrence: Ubiquitous. Found in plutonic rocks, pegmatites, pneumatolytic deposits, exhalative and sedimentary ores as well as in metamorphic environments. Also found in stony meteorites.

Typical occurrences are: Evje (Norway), Sudbury (Ontario, Canada), Aswan (Egypt), Cornwall (UK).

Etch tests: HCl: a large drop causes tarnishing and the solution turns yellow.
Aqua **regia:** solution turns yellow.

MANGANITE

MnO(OH)   Monoclinic

Colour: Black to blackish brown.

Streak: Reddish brown.

Lustre: Submetallic.

Colour **In air:**
Generally: grey to brownish grey.

In oil:
Generally: colours are paler than in air.

→ Pyrolusite: darker grey.

→ Pyrolusite: _____

Bireflectance/pleochroism: Variable. Weak in cross-sections, strong in parallel sections (especially in oil).
// a dark grey with a brown tint.
// b (darkest) with an olive tint.
// c (lightest) light grey-brown.
} Colours noted in oil.

Crossed polars anisotropy: Strong (parallel elongation). May show yellow, blue, grey, violet-grey colours.

Internal reflections: Very common and invariably blood-red (better seen in oil).

Textures: Distinct cleavage. Invariably shows alteration (mainly along the cleavages) to pyrolusite. Prismatic and lamellar crystals are often twinned with a radiating habit.

Associated minerals: Pyrolusite and all the other Mn ores.

Other remarks: None.

Density (g/cm³): 4.3

Reflectance (%) in air:

	at 550 nm	at 580 nm
R_p:	14.8	14.7
R_m:	17.0	17.0
R_g:	21.4	20.7

Mean: 17.5; range: 14.0–22.0

Hardness (Mohs'): 4.0

Hardness (Talmage/polishing): **E+** (< magnetite, ≪ pyrolusite).

VHN: Mean: 565; range: 367–766.

Diagnostic features: Internal reflections (blood-red) and the strong anisotropy, together with the good cleavage and association with other Mn ores, are distinctive.

Occurrence: Restricted to a few low-temperature deposits, mainly of sedimentary origin. Typical occurrences are: Ilfeld (Harz, E. Germany), Anarek (Iran).

Etch tests: HCl: tarnishes and the surface stains light-brown.
H_2O_2: effervesces slowly but does not stain the surface.

MARCASITE

FeS_2 Orthorhombic dimorph of pyrite
(White Iron Pyrites) (Ideal formula)

Colour: Pale bronze-yellow with a green hue but may appear darker due to tendency to tarnish.

Streak: Greyish to brownish black.

Lustre: Metallic.

Colour

In air:	In oil:
Generally: yellowish white with green and pink tints.	Generally: no change.
→ Pyrite: whiter.	→ Pyrite: increased contrast.
→ Arsenopyrite: greenish yellow.	→ Arsenopyrite: increased contrast.
→ Silver: grey and blue tints.	→ Silver: _____

Bireflectance/pleochroism: Strong.
// a: white with a brownish tint, i.e. very similar to pyrite.
// b and c: yellow with a green tint.

Crossed polars anisotropy: Strong and especially so in sections parallel to {100} or {010} when blue, green-yellow or purple and violet-grey colours may be noted.

Internal reflections: Not present.

Textures: Lamellar twins are very common as are arrow-head or four-sector twins. Colloform aggregates with a radiating texture often associated with pyrite are frequent.

Associated minerals: Pyrite, galena, sphalerite, magnetite, pyrrhotite.

Other remarks: None.

Density (g/cm³): 4.85–4.90

Reflectance (%) in air:

at 540 nm	at 580 nm
47.5–56.3	48.4–54.6
Mean: 51.8; range: 48.9–55.5	

Hardness (Mohs'): 6.0–6.5

Hardness (Talmage/polishing):   **E+** (> pyrrhotite, < pyrite).

VHN: Mean: 1121; range: 824–1288.

Diagnostic features: Strong anisotropism (blue to yellow) and high reflectance are useful indicators of marcasite. Can be mistaken for pyrite and arsenopyrite, but the characteristic twinning, hardness, lustre and its paragenetic sequence should help to identify this mineral.

Occurrence: Low-temperature mineral found in Pb-Zn deposits of sedimentary or hydrothermal origin. Formed in the oxidation zone of ore deposits from the alteration of pre-existing minerals.
Typical occurrences are: Sudbury (Ontario, Canada), Cyprus, Joplin (Missouri, USA), Dux (Czechoslovakia), Lancelet (Queensland, Australia), Mt Bischoff (Tasmania).

Etch tests: HNO_3:   stains brown to iridescent; sometimes slow effervescence.

MILLERITE

NiS
(Capillary Pyrites) (Co may substitute for Ni)

Colour:	Pure yellow to brass, pale yellow, sometimes grey with an iridescent tarnish.
Streak:	Greenish black.
Lustre:	Metallic.

Colour	*In air:*	*In oil*:
	Generally: yellow to cream tint.	Generally: no change.
	→ Chalcopyrite: lighter.	→ Chalcopyrite: greenish yellow.
	→ Pentlandite: yellower (no brown tint).	→ Pentlandite: about the same.
	→ Pyrite: slightly yellower.	→ Pyrite: about the same.

Bireflectance/pleochroism:	Distinct with bright yellow to greyish yellow colour (best observed in oil).
Crossed polars anisotropy:	Strong and showing straw or lemon-yellow to an iris-blue or violet colour. It has straight but no complete extinction. Basal sections may appear isotropic.
Internal reflections:	Not present.
Textures:	Often occurs in acicular crystals in a radiating or subparallel aggregate. Large masses may show double twin lamellae. Zonal texture often noted in basal sections.
Associated minerals:	Pentlandite and other Ni minerals, sphalerite, pyrite, hematite, magnetite.
Other remarks:	Often noted as a replacement or alteration phase.
Density (g/cm³):	5.3–5.6

Reflectance (%) in air:

	at 546 nm	*at 589 nm*
R_o:	51.5	53.2
R_e:	56.2	59.0

Mean: 55.4; range: 51.8–60.0

Hardness (Mohs'):	3.0–3.5
Hardness (Talmage/polishing):	E (> chalcopyrite, < sphalerite, pentlandite).
VHN:	Mean: 281; range: 192–383.
Diagnostic features:	The intensity of the colour as well as the polarisation colours (bright-yellow tints) are the most useful aids to identification. The hardness and the unusual shape of the crystals, if present, may also be used.
Occurrence:	It results from the weathering of Ni-rich sulphides and is hence located in the transition between the weathering and cementation zones. Hypogene in origin. Typical occurrences are: Sudbury (Ontario, Canada), Heazlewood (Tasmania).
Etch tests:	HNO_3: effervesces slowly and stains brown.
	$HgCl_2$: may slowly stain brown.

MOLYBDENITE

MoS_2 Trigonal

Colour: Pure lead-grey.

Streak: Bluish grey (**on paper!**), greenish grey on unglazed porcelain.

Lustre: Metallic.

Colour	*In air:*	*In oil:*
	Generally: white (cf. galena).	Generally: strongly decreased. 'e' and 'o' unaltered.
	→ Galena: very similar.	→ Galena: no change.
	→ Graphite: brighter white, no brown tint.	→ Graphite: no change.

Bireflectance/pleochroism: Extremely strong and very characteristic indeed.
'e' – white.
'o' – dull grey with bluish tint.

Crossed polars anisotropy: Very strong indeed. With polars just uncrossed will give a dark-blue colour while with polars at 45° is white with a pink tint.

Internal reflections: Not present.

Textures: Consistently develops lamellar (twisted) texture with a perfect cleavage. May occur as cryptocrystalline aggregates.

Associated minerals: Cassiterite, wolframite, bismuthinite, sphalerite, galena, cobaltite, magnetite, cubanite, Co–Ni arsenides.

Other remarks: Has a greasy feel.

Density (g/cm³): 4.7–4.8

Reflectance (%) in air:

	at 546 nm	*at 589 nm*
R_o:	40.4	38.8
R_e:	19.5	19.0

Mean: 32.4; range: 15.0–44.9

Hardness (Mohs'): 1.0–1.5

Hardness (Talmage/polishing): **B+** (> chalcopyrite, < graphite).

VHN: Mean: 32; range: 16–74.

Diagnostic features: Low hardness, cleavage, the pink-white colour under crossed polars, the lamellar texture and the strong pleochroism are all distinguishing characteristics. The additional test of setting the polars at 45° is useful.

Occurrence: Generally occurs in the pegmatatic–pneumatolytic suite of intrusive sequences as high-temperature veins. Also noted from Au placers. Typical occurrences are: Climax (Colorado, USA), Tunaberg (Sweden), Azegour (Morocco).

Etch tests: Negative to all reagents.

NICCOLITE

NiAs   Hexagonal
(Nickeline, Copper Nickel) (Fe and Co can replace Ni; miscible with NiSb)

Colour: Pale-copper-red.

Streak: Pale-brownish black.

Lustre: Metallic.

Colour

In air:	*In oil*:
Generally: yellowish pink.	Generally: no change.
→ Pyrrhotite: more pinkish.	→ Pyrrhotite: even more pink.
→ Bornite: lighter in colour.	→ Bornite: even lighter.
→ Cobaltite: more coloured.	→ Cobaltite: lighter.

Bireflectance/pleochroism: Strong.
'o' – yellow-pink/light pink.
'e' – brownish pink.

Crossed polars anisotropy: Very strong with yellow, greenish yellow and violet-blue colours. It has straight extinction which is never complete. Basal sections appear isotropic.

Internal reflections: Not present.

Textures: Aggregates of radiating crystals are common, as are xenomorphic crystals. A chevron-like texture of isotropic and anisotropic lamellae, possibly due to twinning, is frequently seen.

Associated minerals: Molybdenite, gold, pitchblende, Ag and Bi minerals, Co–Ni minerals.

Other remarks: May form pseudo-eutectic intergrowths with pyrrhotite, chalcopyrite and other minerals.

Density (g/cm³):   7.7–7.8

Reflectance (%) in air:

at 546 nm	*at 589 nm*
46.4–51.6	53.1–56.0
Mean: 55.3; range: 52.0–58.3	

Hardness (Mohs'):   5.0–5.5

Hardness (Talmage/polishing):   E (≫ silver, > chalcopyrite, ∼ pyrrhotite, ≪ löllingite, pyrite).

VHN: Mean: 419; range: 308–533.

Diagnostic features: The reflectance (orange colour) and the anisotropy are characteristic. Pyrrhotite, with which it may be confused, is less reflective and more yellow-brown.

Occurrence: Abundant worldwide as an intrusive–hydrothermal vein mineral; may also occur in basic igneous rocks. Typical occurrences are: Cobalt (Ontario, Canada), Schneeberg and Annaberg (E. Germany), Talmessi (Iran), Cusco (Peru).

Etch tests: HNO_3:   effervesces and tarnishes.
$FeCl_3$:   stains brown.
$HgCl_2$:   sometimes stains brown.

PENTLANDITE

$(Fe, Ni)_9S_8$ Cubic
(Co can replace Ni totally; Ag content up to 4%)

Colour: Light bronze-yellow.

Streak: Light bronze-brown.

Lustre: Metallic.

Colour

In air:	*In oil:*
Generally: creamish white tint may be pink to brown.	Generally: no change.
→ Pyrrhotite: brighter and no pink/brown tint.	→ Pyrrhotite: brighter and creamy white.

Bireflectance/pleochroism: Not present.

Crossed polars anisotropy: Isotropic.

Internal reflections: Not present.

Textures: Diagnostic octahedral cleavage. Often occurs as **minute** exsolution 'flames' in pyrrhotite and may also be replaced by pyrrhotite along cleavages. It alters to millerite along cracks and grain boundaries.

Associated minerals: Pyrrhotite, chalcopyrite, cubanite, magnetite, ilmenite, chromite and the Pt minerals.

Other remarks: It is **not** magnetic.

Density (g/cm³): 4.96 (approx.)

Reflectance (%) in air:

at 546 nm	at 589 nm
46.5–50.5	49.0–52.0
Mean: 51.0; range: 49.6–52.0	

Hardness (Mohs'): 3.5–4.0

Hardness (Talmage/polishing): **D−** (> chalcopyrite, < pyrrhotite).

VHN: Mean: 220; range: 195–238.

Diagnostic features: The paragenetic association with pyrrhotite and chalcopyrite is very useful. Its octahedral cleavage and possible exsolution textures (use high magnification) are distinctive. It may be confused with pyrite (if in small grains) but a check on the polishing hardness normally differentiates the two minerals.

Occurrence: Magmatic differentiates in gabbros and norites.
Typical occurrences are: Sudbury (Ontario, Canada), Monchegorsk (USSR), Outokumpu (Finland), Insizwa (S. Africa), Kambalda (W. Australia).

Etch tests: HNO_3: tarnishes and slowly stains brown.

PLATINUM

Pt   Cubic
(Always contains Fe up to 20%, may contain Ir, Cu, Pd, Rh, Ru etc.)

Colour: White to steel-grey, often said to be tin-white.

Streak: Whitish steel-grey.

Lustre: Metallic.

Colour   *In air:*   *In oil*:
            Generally: bluish white.   Generally: no change.
            → Silver: whiter.   → Silver: no change.

Bireflectance/pleochroism: Not present.

Crossed polars anisotropy: Isotropic with incomplete extinction.

Internal reflections: Not present.

Textures: Xenomorphic polygonal grains are frequently noted, as is a zonal texture. Exsolution bodies of iridium are common and it may contain inclusions of chromite, hematite, etc.

Associated minerals: Pentlandite and the rare platinoids, chromite, magnetite.

Other remarks: Sometimes magnetic.

Density (g/cm³):   14.0–19.0 (if pure 21.0–22.0)

Reflectance (%) in air:

at 546 nm	*at 589 nm*
70.3	71.9
Mean: 70.0	

Hardness (Mohs'):   4.0–4.5

Hardness (Talmage/polishing):   _____ (> sphalerite, < pyrrhotite).

VHN: Mean: 128; range: 114–146.

Diagnostic features: The very high reflectance, hardness and association with ultramafic rocks (sometimes with other platinoids) are diagnostic. It may be confused with palladium – but see test with HCl and chromic acid mixture.

Occurrence: Rare, but with economic proportions in a few localities. Magmatic segregation deposits and some placer deposits as well as hydrothermal veins.
Typical occurrences are: Bushveld (S. Africa), Sudbury (Ontario, Canada), the Urals (USSR).

Etch tests: Negative to all **normal** reagents.
Aqua regia: dissolves Pt.
Chromic acid + HCl: dissolves Pt.

PROUSTITE

Ag_3AsS_3
(Light Ruby Silver) ($3Ag_2S.As_2S_3$)

Colour:	Scarlet-vermilion.
Streak:	Scarlet-vermilion.
Lustre:	Adamantine.

Colour	*In air:*	*In oil:*
	Generally: grey to blue-white.	Generally: distinctly lower masked by internal reflections.
	→ Pyrargyrite: darker and bluer.	→ Pyrargyrite: more distinct.

Bireflectance/pleochroism: Distinct.
In air: 'o' – white with yellow tint.
'e' – bluish grey and darker.
In oil: 'o' – greyish blue with a brown tint.
'e' – dark greyish blue.

Crossed polars anisotropy: Strong, especially in oil, but masked by internal reflections.

Internal reflections: Always present. Intense (more so than pyrargyrite) scarlet-red.

Textures: Irregular grains and allotriomorphic aggregates, or idiomorphic needle-like crystals. Many samples are twinned and zoned.

Associated minerals: Chalcopyrite, arsenopyrite, tetrahedrite, Co–Ni–Fe arsenides, galena, silver.

Other remarks: cf. pyrargyrite.

Density (g/cm³): 5.57–5.64 (5.57 if pure).

Reflectance (%) in air:

at 546 nm	*at 589 nm*
27.4–28.1	25.8–26.4

Mean: 27.3; range: 25.0–28.4

Hardness (Mohs'): 2.0–2.5

Hardness (Talmage/polishing): **B−** (∼ pyrargyrite).

VHN: Mean: 119; range: 90–143.

Diagnostic features: It is **very** difficult to distinguish proustite from pyrargyrite. The streak is possibly the only indicator (scarlet-vermilion or brick-red).

Occurrence: Infrequent. Occurs in supergene and hypogene deposits. It is rarer and more sparse than pyrargyrite.
Typical occurrences are: Cobalt (Ontario, Canada), Comstock (Nevada, USA), Ruby District (Colorado, USA) and numerous localities in Bolivia.

Etch tests: HNO_3:   tarnishes sometimes.
KCN: stains black and brings out possible scratches.

FeCl$_3$:   may stain grey to black.
KOH: stains grey to black.
HgCl$_2$:   stains brown.

PSILOMELANE

Orthorhombic or Monoclinic
(no general formula as name is given to all massive hard manganese minerals)

Colour: Iron-black to dark-steel-grey.

Streak: Brownish black.

Lustre: Submetallic or dull.

Colour	*In air:*	*In oil:*
	Generally: blue-grey to grey-white	Generally: minor variations.
	→ Pyrolusite: darker.	→ Pyrolusite: _____
	→ Magnetite: bluish white.	→ Magnetite: _____

Bireflectance/pleochroism: Strong. // c – white (even in oil).
          ⊥ c – grey to blue-grey.

Crossed polars anisotropy: Strong. White to grey with straight extinction.

Internal reflections: Infrequent; if present, are brown.

Textures: Cryptocrystalline and amorphous textures as well as colloform, concentric and spherulitic textures are noted.

Associated minerals: All manganese minerals, goethite and other Fe hydroxides.

Other remarks: None

Density (g/cm^3): 4.5–4.7

Reflectance (%) in air:

at 546 nm	*at 589 nm*
15	30

Mean: 23.9; range: 23.0–24.4

Hardness (Mohs'): 5.0–7.0

Hardness (Talmage/polishing): **D−** (< coarser-grained specimens than other Mn minerals).

VHN: Mean: 667; range: 503–813.

Diagnostic features: Colloform and other similar textures; moderate reflectance. It may require X-ray diffraction to confirm its presence.

Occurrence: Widespread geographically and geologically. A surface product in the sedimentary cycle and as such formed at low temperatures and pressures in atmospheric conditions.

Etch tests: HNO$_3$:   stains light brown, fumes may tarnish.
          HCl: stains brown to black.
          FeCl$_3$:   may stain light brown.
          H$_2$O$_2$:   effervesces, no etching.

PYRARGYRITE

Ag_3SbS_3
(Dark Ruby Silver) $(3Ag_2S.Sb_2S_3)$

Colour:	Grey-black to bluish grey.
Streak:	Purplish red.
Lustre:	Adamantine to metallic.

Colour

In air:	*In oil:*
Generally: bluish grey.	Generally: strongly lowered, more grey-blue.
→ Proustite: white to light cream.	→ Proustite: lighter.
→ Galena: pale blue/grey-white.	→ Galena: grey-blue.

Bireflectance/pleochroism:	Distinct to strong. No marked colour variations.
Crossed polars anisotropy:	Strong. Masked by the internal reflections, if viewed in oil. In air may be pale-grey to dark-grey.
Internal reflections:	Intense. Carmine-red, but less pronounced than proustite.
Textures:	Idiomorphic crystals and aggregates of irregular grains are common. Twins and zonal textures are often noted. It may contain inclusions or occur as inclusions.
Associated minerals:	Argentite, silver, galena, chalcopyrite, tetrahedrite, bournonite, boulangerite, stibnite, proustite.
Other remarks:	None.
Density (g/cm³):	5.77–5.86

Reflectance (%) in air:

	at 546 nm	at 589 nm
R_o:	30.3	28.4
R_e:	28.5	26.5

Mean: 30.2; range: 28.4–30.8

Hardness (Mohs'):	2.0–2.5
Hardness (Talmage/polishing):	C (≫ argentite, < galena).
VHN:	Mean: 100; range: 50–127.
Diagnostic features:	Can be confused with proustite and it is advisable therefore to test the streak.
Occurrence:	Widespread in hydrothermal veins – more so than proustite. Common in Pb–Zn–Ag and Ag–Co–Ni veins. Paragenetically it is a late-formed mineral and can therefore be of supergene or hypogene origin, and thus often associated with cementation textures.
	Typical occurrences are: Broken Hill (NSW, Australia), Kongsberg (Norway), Bolivian deposits, Cobalt (Ontario, Canada), Freiberg (E. Germany).

Etch tests:

KCN:	light brown stain and possible black tarnish.
KOH:	iridescent and then a black tarnish.
$HgCl_2$:	may stain brown.
HI:	structurally etches.

PYRITE

FeS$_2$ Cubic
(Iron Pyrites) (May contain Co, Ni, As and ? Au, Cu up to 10 wt%)

Colour:	Light yellow to brass-yellow.
Streak:	Greenish black to brownish black.
Lustre:	Strongly metallic (splendent).

Colour	*In air:*	*In oil:*
	Generally: whitish yellow.	Generally: no change.
	→ Marcasite: yellower.	→ Marcasite: even yellower.
	→ Arsenopyrite: cream-yellow.	→ Arsenopyrite: no change.
	→ Galena: pale-yellow.	→ Galena: _____
	→ Silver: greyer and greenish.	→ Silver: _____
	→ Chalcopyrite: pale-yellow.	→ Chalcopyrite: yellow to deep-yellow.

Bireflectance/pleochroism:	Not present.
Crossed polars anisotropy:	Weak to distinctly anisotropic (blue-green to orange-red). May vary depending on how it has been polished and whether or not lattice distortions are present.
Internal reflections:	Not present.
Textures:	Mainly forms idiomorphic crystals. Zoning is common as is a cellular texture. Spherical aggregates, more often called framboids, are also developed.
Associated minerals:	Galena, sphalerite, pyrrhotite, chalcopyrite, arsenopyrite, marcasite, Pb sulphosalts.
Other remarks:	None.
Density (g/cm^3):	4.95–5.10
Reflectance (%) in air:	at 546 nm at 589 nm 51.7–53.4 53.4–55.0 Mean: 54.5; range: 54.1–54.8
Hardness (Mohs'):	6.0–6.5
Hardness (Talmage/polishing):	F (> arsenopyrite, marcasite, hematite; < cassiterite).
VHN:	Mean: 1377; range: 1027–1836.
Diagnostic features:	The yellow-white colour, high reflectance and the polishing hardness are all diagnostic.
Occurrence:	Ubiquitous in: granites, pegmatites, hydrothermal deposits (low and high temperature), sedimentary and metamorphic environments. Typical occurrences are: Rio Tinto (Spain), Sudbury (Ontario, Canada), Kleva and Kiruna (Sweden), Merensky Reef (Rustenburg) and the Witwatersrand (S. Africa), Mt Bischoff (Tasmania, Australia), numerous localities in Norway, Butte (Montana) and Cripple Creek (Colorado, USA), as well as other localities too numerous to mention throughout the world.
Etch tests:	HNO$_3$: slowly effervesces and stains iridescent.

PYROLUSITE

MnO_2 Tetragonal
(Polianite) (Name often restricted to the soft Mn oxides)

Colour: Iron-black, dark steel-grey and sometimes bluish.

Streak: Black or bluish black.

Lustre: Metallic.

Colour

In air:	*In oil:*
Generally: white with creamy tint.	Generally: more pronounced.
$\rightarrow$ Magnetite: yellowish white.	$\rightarrow$ Magnetite: more pronounced.
$\rightarrow$ Hematite: yellowish.	$\rightarrow$ Hematite: more pronounced.
$\rightarrow$ Manganite: pure white.	$\rightarrow$ Manganite: more pronounced.

Bireflectance/pleochroism: Distinct and especially so for coarse-grained samples in oil.
// e – yellowish white.
// o – white to grey.

Crossed polars anisotropy: Very strong. Yellow to dark-brown, greenish blue or slate-grey if the polars are perfectly crossed. Fine-grained aggregates appear isotropic. Straight extinction.

Internal reflections: Not present.

Textures: Cryptocrystalline aggregates or coarse-grained euhedral tabular or prismatic crystals. Twinning, if present, may be single or lamellar.

Associated minerals: Goethite, manganite, magnetite and all the other Mn oxides.

Other remarks: None.

Density (g/cm^3): 4.73–4.86

Reflectance (%) in air:

at 546 nm	*at 589 nm*
29.0–40.0	28.1–39.3

Mean: 34.4; range: 30.0–41.5

Hardness (Mohs'): 1.0–2.0 (massive)
6.0–6.5 (crystalline)

Hardness (Talmage/polishing): **D to E** (highly variable and depending on the type of aggregate).

VHN: Mean: 266; range: 225–405.

Diagnostic features: The paragenetic association with other Mn ores, its hardness, high reflectance and brownish yellow polarisation colours.

Occurrence: Ubiquitous.
Typical occurrences are: Ilfeld (Harz, E. Germany), Imini (Morocco).

Etch tests: HCl: brown colour, may stain.
Aqua regia: as for HCl.
$H_2O_2 + H_2SO_4$: stains black.

PYRRHOTITE (GROUP)

$Fe_{1-x}S$ [~Fe_9S_{10}-Hexagonal]
(Magnetic Pyrites) (may contain Ni, Co and Mn) [~Fe_7S_8-Monoclinic]
(Pyrrhotine)

Colour:	Varies from bronze-yellow to copper-red. Tarnishes easily.
Streak:	Greyish black.
Lustre:	Metallic.

Colour	*In air:*	*In oil:*
	Generally: cream with faint pinkish brown tint.	Generally: lowered, colour now more brown than pink.
	→ Pentlandite: darker.	→ Pentlandite: more distinct.
	→ Cubanite: pinkish.	→ Cubanite: more distinct.
	→ Niccolite: darker, no reddish tints.	→ Niccolite: more distinct.

Bireflectance/pleochroism:	Very distinct. 'e' – brownish cream. 'o' – reddish brown.
Crossed polars anisotropy:	Very strong. Yellow-grey, greyish green or greyish blue. The intensity depends on the orientation.
Internal reflections:	Not present.
Textures:	(Seldom euhedral.) Deformation twins common and flame exsolution bodies of pyrrhotite in pentlandite are frequent. Mixture of pyrrhotite + pyrite + marcasite (supergene alteration) may give concentric bird's eye textures. It is often replaced.
Associated minerals:	Pyrite, marcasite, magnetite, chalcopyrite, cubanite, pentlandite, ilmenite, galena, arsenopyrite, hematite.
Other remarks:	Magnetic. Stoichiometric FeS is called troilite.
Density (g/cm³):	4.58–4.64

Reflectance (%) in air

		at 546 nm	*at 589 nm*
Hexagonal:	R_o:	34.0	35.8
	R_e:	39.2	40.7
Monoclinic:	R_o:	34.8	36.9
	R_e:	39.9	41.6

Mean: 39.3; range: 34.0–45.6

Hardness (Mohs'):	3.5–4.5
Hardness (Talmage/polishing):	**D**− (≫ chalcopyrite, niccolite, ≪ arsenopyrite, pyrite).
VHN:	Mean: 278; range: 212–363.

Diagnostic features:	The light brown colour, strong anisotropism, pleochroism (in oil) and its hardness. Pentlandite is isotropic and cubanite is softer and anisotropic.
Occurrence:	Ubiquitous. A high-temperature mineral of varying origins: magmatic associations, pegmatites, hydrothermal deposits and meteorites. Typical occurrences are: Bushveld (S. Africa), Rammelsberg (W. Germany), Bolivian deposits.
Etch tests:	HNO_3: stains light brown. HCl: drop turns yellow but surface not stained. KOH: slowly turns brown near edge of drop.

RUTILE

TiO_2 Tetragonal
(May contain Fe, Nb, Ta, Sn and Cr in solid solution)

Colour:	Reddish brown; may be green, purple or colourless.
Streak:	Pale-brown.
Lustre:	Metallic, adamantine.

Colour	*In air:*	*In oil:*
	Generally: grey with bluish tint.	Generally: pale grey.
	→ Chromite: no change.	→ Chromite: _____
	→ Magnetite: no change.	→ Magnetite: _____
	→ Ilmenite: no brown tint.	→ Ilmenite: _____
	→ Hematite: pinkish blue.	→ Hematite: no change.
	→ Goethite: pinkish blue.	→ Goethite: _____
	→ Cassiterite: much lighter.	→ Cassiterite: _____

Bireflectance/pleochroism:	Distinct and especially so in oil. ⊥ c ('o') – lightest.
Crossed polars anisotropy:	Strong. The colours often masked by internal reflections. (Nb–rutile is weakly anisotropic.)
Internal reflections:	Abundant and strongly coloured. The colours may be white, yellow, brown, red-brown, violet, green etc., depending on the elements present. Better seen in oil.
Textures:	Polysynthetic twins are common as is cataclastic texture and zoning, which is often noted in thin sections, but is **not** seen in polished samples. It has an idiomorphic tendency.
Associated minerals:	Hematite, ilmenite, magnetite, quartz.
Other remarks:	None.
Density (g/cm³):	4.23–4.40

Reflectance (%) in air:

	at 546 nm	at 589 nm
R_o:	19.8–20.0	19.4–19.6
R_e:	23.0–23.5	22.3–23.0

Mean: 22.4; range: 20.0–24.6

Hardness (Mohs'): 6.0–6.5

Hardness (Talmage/polishing): **G**± (> ilmenite, < hematite and cassiterite).

VHN: Mean: 1046; range: 978–1280.

Diagnostic features: The internal reflections, twins, hardness and the high reflectance are all characteristic. Cassiterite is similar but has a lower reflectance and is more difficult to polish.

Occurrence: Ubiquitous. It ranges throughout all depositional environments, i.e. igneous, sedimentary and metamorphic.
Typical occurrences are: Boliden (Sweden), Bor (Yugoslavia), Witwatersrand (S. Africa).

Etch tests: Negative to all reagents.

SCHEELITE

$CaWO_4$ Tetragonal
(Mo may substitute for W)

Colour:	White, colourless or yellow, brown, green or red.
Streak:	White.
Lustre:	Vitreous to adamantine.

Colour	*In air:*	*In oil:*
	Generally: grey-white.	Generally: very dark grey.
	→ Chalcopyrite: dark grey, sometimes violet tint.	→ Chalcopyrite: dark grey.
	→ Arsenopyrite: very dark grey.	→ Arsenopyrite: dark grey.

Bireflectance/pleochroism:	Not present.
Crossed polars anisotropy:	Distinct, but masked by the internal reflections.
Internal reflections:	Abundant, white.
Textures:	Usually coarse-grained masses which are rarely idiomorphic. Often noted replacing wolframite.
Associated minerals:	Wolframite, arsenopyrite, pyrrhotite, bismuthinite, molybdenite and most of the 'skarn' minerals.
Other remarks:	None.
Density (g/cm³):	6.0

Reflectance (%) in air:

	at 546 nm	*at 589 nm*
R_o:	10.0	9.9
R_e:	10.3	10.1
Mean: 11.0; range: 10.0–12.1		

Hardness (Mohs'):	5.0
Hardness (Talmage/polishing):	**D−** (< wolframite, cassiterite).
VHN:	Mean: 361; range: 285–464.
Diagnostic features:	(May be mistaken for zircon.) Under UV light it fluoresces pale-blue to yellow.
Occurrence:	Widespread, but mainly with tungsten deposits (contact pneumatolytic). Typical occurrences are: Adelong (NSW, Australia), Sonora (Mexico), Mt Ramsay (Tasmania, Australia), Meymac (France) and Carrock Fell (Cumbria, UK).
Etch tests:	Negative to all reagents.

SILVER

Ag   Cubic
(May contain Au, Hg, As, Sb, Pt, Ni, Pb, Fe, Cu and Bi)

Colour:	Bright white (silver-white). Rapidly tarnishes to cream or brown, grey or black. Often shows iridescence.
Streak:	Silver-white.
Lustre:	Metallic.

Colour	*In air:*	*In oil:*
	Generally: bright white with a creamy tint.	Generally: no change.
	→ Copper: lighter white.	→ Copper: the same.
	→ Platinum: creamy white and even lighter.	→ Platinum: the same.

Bireflectance/pleochroism:	Not present.
Crossed polars anisotropy:	Isotropic, but sometimes is scratched and the pits may give false light effects.
Internal reflections:	Not present.
Textures:	As inclusions in argentite, but mainly as dendritic or skeletal inclusions. Twinning and zoning not uncommon.
Associated minerals:	All Ag minerals, Ni–Co arsenides, sphalerite, calcite, galena, tetrahedrite–tennantite.
Other remarks:	White silver-gold alloys are called Electrum.
Density (g/cm^3):	10.1–11.1

Reflectance (%) in air:

at 546 nm	*at 589 nm*
94.2–94.8	95.0–95.5
Mean: 95.1; range: 93.8–96.8	

Hardness (Mohs'):	2.5–3.0
Hardness (Talmage/polishing):	**B** (≫ proustite, > galena, < tetrahedrite, ≪ sphalerite).
VHN:	Mean: 58; range: 39–118.
Diagnostic features:	Extremely high reflectance, hardness and its ability to tarnish easily are all diagnostic.
Occurrence:	Abundant in hydrothermal solutions (epithermal) and sedimentary formations (in the oxidising and enrichment zones) with either limonite and/or argentite. Typical occurrences are: Cobalt (Ontario, Canada), Freiberg (E. Germany), Kongsberg (Norway).
Etch tests:	HNO_3: effervesces and leaves a white coating.
	HCl: sometimes tarnishes.
	KCN: may tarnish grey to brown.
	$FeCl_3$: **instant** iridescence.
	$HgCl_2$: stains brown with or without iridescence.

SPHALERITE

(Zn,Fe)S   Cubic
(Zinc Blende, Black Jack, Blende, False Galena) (May contain Mn, Cd, Ga, In, Ti and Hg)

Colour:	Colourless to very dark brown; may be yellow, red or green.
Streak:	Brown to light yellow or white.
Lustre:	Adamantine or resinous.

Colour	*In air:*	*In oil:*
	Generally: grey with brown tint.	Generally: very dark grey.
	→ Magnetite: darker.	→ Magnetite: much darker.

Bireflectance/pleochroism:	Not present.
Crossed polars anisotropy:	Isotropic, but probably due to deformations or chemical admixtures may appear anisotropic, when shades of grey are seen.
Internal reflections:	Always noted, especially in oil. Fe-rich specimens: reddish brown. Fe-poor specimens: yellowish brown.
Textures:	Exsolution textures are numerous, as are coarse granular aggregates and idiomorphic crystals. Zoning (often seen with the naked eye) and glide (translation) twinning are also frequent.
Associated minerals:	Galena, chalcopyrite, pyrite, stannite, arsenopyrite, pyrrhotite, magnetite.
Other remarks:	Colloform variety is called 'schalenblende'. Iron-rich sphalerite and wurtzite is known as marmatite.
Density (g/cm³):	3.9–4.08 (pure)

Reflectance (%) in air:

at 546 nm	*at 589 nm*
16.1–22.2	15.8–19.0
Mean: 17.3; range: 16.1–18.8	

Hardness (Mohs'):	3.5–4.0
Hardness (Talmage/polishing):	C− (> chalcopyrite, tetrahedrite, stannite, enargite; < pyrrhotite, magnetite and ilmenite).
VHN:	Mean: 199; range: 128–276.
Diagnostic features:	Low reflectance, which decreases even further in oil, the internal reflections and its hardness are all diagnostic. Magnetite is browner, harder and has no internal reflections.
Occurrence:	Ubiquitous: magmatic, pneumatolytic, contact metamorphic, hydrothermal (veins etc.), sedimentary and metamorphic environments. Typical occurrences are : Sudbury (Ontario, Canada), Tri-State (USA), Pennines (UK), Butte (Montana, USA), Broken Hill (NSW, Australia).
Etch tests:	HNO_3: tarnishes with or without effervescence and indicates twinning (if present).

HCl: tarnishes.
Aqua regia: effervesces and stains dark brown.

STANNITE

Cu_2FeSnS_4 Tetragonal
(Tin Pyrites) (Fe replaced by Zn)

Colour: Steel-grey to iron-black, sometimes with bluish tarnish; if next to chalcopyrite, appears to be yellow.

Streak: Black.

Lustre: Metallic.

Colour

In air:	In oil:
Generally: brownish olive-grey.	Generally: _____
→ Tetrahedrite: darker, brown-grey.	→ Tetrahedrite: _____
→ Sphalerite: lighter, yellow-brown.	→ Sphalerite: _____
→ Chalcopyrite: much darker.	→ Chalcopyrite: _____
→ Cassiterite: brownish olive-green.	→ Cassiterite: _____

Bireflectance/pleochroism: Distinct. Light brown to brownish olive-green.

Crossed polars anisotropy: Strong. Yellowish brown, greyish olive-green, bluish or violet-grey colours.

Internal reflections: Not present.

Textures: Orthogonal crystals and twins produce a microcline-like appearance. Exsolution of chalcopyrite, or in chalcopyrite, often noted. Lamellar or coarse granular aggregates commonly occur as space fillings.

Associated minerals: Arsenopyrite, cassiterite, wolframite, chalcopyrite, sphalerite.

Other remarks: For further reading and more detailed information, see Ramdohr (1980: 548).

Density (g/cm³): 4.3–4.5

Reflectance (%) in air:

	at 546 nm	at 589 nm
R_o:	26.0	26.1
R_e:	27.8	27.7

Mean: 27.6; range: 27.1–28.0

Hardness (Mohs'): 4.0

Hardness (Talmage/polishing): **D+** (> chalcopyrite, ~ tetrahedrite, < sphalerite).

VHN: Mean: 216; range: 171–307.

Diagnostic features: The polarisation tints (especially violet), the colour, twinning and its paragenesis aid in recognition.

Occurrence: Abundant in high-temperature hydrothermal deposits genetically related to cassiterite veins.

Typical occurrences are: Cornwall (UK), Aberfoyle (Tasmania), Ashio (Japan), Lutzi Ngau (Hunan, China), various localities in Bolivia.

Etch tests: HNO_3: stains iridescent to black.

STIBNITE

Sb_2S_3 Orthorhombic
(Antimonite, Antimony Glance)

Colour: Lead-grey to steel-grey. May tarnish black and iridescent.

Streak: Lead-grey.

Lustre: Metallic (splendent on cleavage surfaces).

Colour	*In air:*	*In oil:*
	Generally: white to greyish white.	Generally: decreased.
	→ Galena: darker.	→ Galena: no change.
	→ Bournonite: lighter.	→ Bournonite: ———
	→ Bismuthinite: darker.	→ Bismuthinite: decreased.

Bireflectance/pleochroism: Strong.
// a – dull grey-white.
// b – brownish white.
// c – pure white.

Crossed polars anisotropy: Very strong. At 45° – blue, grey-white or brown. Ideally straight extinction but undulose extinction is very common.

Internal reflections: Not present.

Textures: Polysynthetic twins (wedge-shaped, and due to pressure) are common. Radiating needle aggregates also noted. Granoblastic textures as seen are growth zones, if the sample is etched.

Associated minerals: Zinkenite, Ag minerals, cinnabar, gold, arsenopyrite.

Other remarks: None.

Density (g/cm^3): 4.52–4.63

Reflectance (%) in air:

at 546 nm	*at 589 nm*
31.1–48.1	30.6–45.2
Mean: 38.0; range: 30.2–47.0	

Hardness (Mohs'): 2.0 (orientation dependent).

Hardness (Talmage/polishing): **B** (< galena, bournonite; ≪ chalcopyrite).

VHN: Mean: 98; range: 42–163.

Diagnostic features:	Its reaction with KOH and low hardness, as well as the pleochroism and anisotropy, all characterise stibnite.
Occurrence:	Widespread as a hydrothermal high- to low-temperature mineral. Replacement deposits are of economic importance. Often formed at less than 100 °C. Typical occurrences are: Shikoku (Japan), Sikwanschan (China), Massiac (Cantal, France).
Etch tests:	HNO_3: stains iridescent and may effervesce. HCl: tarnishes and may stain brown. KCN: stains light brown. KOH: instantly tarnishes brown and gives a yellow coating.

SYLVANITE

$AuAgTe_4$ Monoclinic
(Graphic Tellurium) (May contain excess of Au)

Colour: Silvery white with a yellowish tint.

Streak: Steel-grey.

Lustre: Metallic, brillant.

Colour	*In air:*	*In oil:*
	Generally: creamy-white.	Generally: _____
	→ Galena: lighter.	→ Galena: no change.
	→ Calaverite: darker.	→ Calaverite: considerably darker.
	→ Nagyagite: lighter.	→ Nagyagite: lighter.

Bireflectance/pleochroism: Distinct. Creamy white to creamy brown.

Crossed polars anisotropy: Strong. Light bluish grey to dark brown. No definite extinction. If polars are uncrossed, the tints are more definitive.

Internal reflections: Not present.

Textures: Characteristic texture is the twin lamellae which are noted even in white light due to pleochroism.

Associated minerals: Nagyagite, petzite, calaverite and other tellurides, sphalerite, bornite, chalcopyrite, pyrite.

Other remarks: None.

Density (g/cm³): 7.90–8.3

Reflectance (%) in air:

at 546 nm	*at 589 nm*
50.3–59.0	50.8–59.0
Mean: 55.1; range: 48.0–60.0	

Hardness (Mohs'): 1.5–2.0

Hardness (Talmage/polishing): **C** (≫ argentite, > nagyagite, < pyrargyrite).

VHN: Mean: 144; range: 102–203.

Diagnostic features: Twin lamellae, high reflectance, strong anisotropism and its paragenesis with other tellurides are useful diagnostic features.

Occurrence: Frequent, as it is the most common of all the Au tellurides. Subvolcanic hydrothermal Ag–Au–Te veins.
Typical occurrences are: Kalgoorlie (W. Australia), Offenbanyia (Roumania).

Etch tests: HNO_3:   slowly effervesces and the surface stains brown.
$FeCl_3$:   may give light yellow stain.
Aqua regia: weak effervescence, may tarnish.

TENNANTITE

$(Cu,Fe)_{12}As_4S_{13}$ Cubic
$(Cu_3AsS_{3.25})$
(Sometimes containing Fe, Zn, Sb etc.)

Colour: Blackish lead-grey to iron-black.

Streak: Reddish brown, black or dark red.

Lustre: Metallic.

Colour	In air:	In oil:
	Generally: grey with bluish green tint.	Generally: _____
	→ Galena: greenish.	→ Galena: _____
	→ Chalcocite: greenish.	→ Chalcocite: _____
	→ Chalcopyrite: bluish grey.	→ Chalcopyrite: _____

Bireflectance/pleochroism: Not present.

Crossed polars anisotropy: Isotropic.

Internal reflections: Frequently noted in shades of red.

Textures: Often occurs as inclusions in galena. Zoning is invariably present as are breccia and cataclastic structures.

Associated minerals: Galena, sphalerite, bournonite, chalcopyrite, Co–Ni minerals and Ag minerals.

Other remarks: With tetrahedrite, known as the fahlore group of minerals.

Density (g/cm³): 4.4–5.1 (and more)

Reflectance (%) in air:

at 546 nm	at 589 nm
30.0	29.8
Mean: 29.1; range: 28.8–29.7	

Hardness (Mohs'): 3.5–4.5

Hardness (Talmage/polishing): **B or D** (≫ bornite, > galena, < enargite and sphalerite).

VHN: Mean: 358; range: 308–401.

Diagnostic features: As for tetrahedrite.

Occurrence: Numerous but predominantly in the hydrothermal suite of ores, e.g. in contact-metamorphic or sub-volcanic veins. Sedimentary (sulphide) ores and metamorphic deposits. Normally in a more As-rich environment than tetrahedrite.
Typical occurrences are: Tsumeb (Namibia, SW Africa), Cornwall (UK), Broken Hill (NSW, Australia), Cobalt (Ontario, Canada), Butte (Montana, USA), and various Bolivian localities.

Etch tests: HNO_3:   tarnishes and may stain iridescent.
           KCN: stains slowly to a light brown colour.

TETRAHEDRITE

$Cu_{12}Sb_4S_{13}$ Cubic
$(Cu_3SbS_{3.25})$
(May contain Fe, Zn, Hg, Bi, Te and Pb)

Colour: Flint-grey to iron-black.

Streak: Grey to black and may be brown or red.

Lustre: Metallic and often splendent.

Colour	*In air:*		*In oil:*	
	Generally: grey or olive with a brownish tint.		Generally: _____	
	→ Galena:	brownish or greenish grey.	→ Galena:	more distinct.
	→ Chalcopyrite:	bluish grey.	→ Chalcopyrite:	_____
	→ Bournonite:	darker brown.	→ Bournonite:	darker or no change.
	→ Stannite:	light-grey.	→ Stannite:	brighter.
	→ Sphalerite:	lighter.	→ Sphalerite:	much brighter.

Bireflectance/pleochroism: Not present.

Crossed polars anisotropy: Isotropic.

Internal reflections: May or may not be noted; if present, they are brownish red.

Textures: Often occurs as inclusions in galena. Zoning is invariably present and sometimes cataclastic and breccia textures.

Associated minerals: Galena, sphalerite, bournonite, chalcopyrite, Co–Ni minerals and Ag minerals.

Other remarks: With tennantite, known as the fahlore group of minerals.

Density (g/cm³): 4.4–5.1 (and more)

Reflectance (%) in air:

at 546 nm	*at 589 nm*
30.5	30.0
Mean: 30.3; range: 29.3–31.2	

Hardness (Mohs'): 3.5–4.5

Hardness (Talmage/polishing): **D** (≫ galena, ∼ chalcopyrite, < sphalerite).

VHN: Mean: 360; range: 291–464.

Diagnostic features: Moderate reflectance and its paragenesis.

Occurrence: Numerous and predominantly in hydrothermal suite of ores, e.g. contact-metasomatic and subvolcanic veins. In sedimentary (sulphide) veins and metamorphic terrains.
Typical occurrences are: Tsumeb (Namibia, SW Africa), Cornwall (UK), Broken Hill (NSW, Australia), Butte (Montana, USA) and Cobalt (Ontario, Canada).

Etch tests: HNO_3: tarnishes and may stain iridescent.
KCN: stains slowly to a light brown colour.

URANINITE

UO$_2$ (not U$_3$O$_8$)　　　　Cubic
(Pitchblende, Nasturan, Uranpecherz) ($\pm$ Th, Pb and the rare earths)

Colour:	Grey, greenish, brownish or velvet-black.
Streak:	Brownish black, grey or olive-green.
Lustre:	Submetallic to greasy or dull.

Colour	*In air:*	*In oil:*
	Generally: grey with brown tint.	Generally: much darker with a distinct brown colour.
	$\rightarrow$ Magnetite: slightly darker.	$\rightarrow$ Magnetite: _____
	$\rightarrow$ Sphalerite: brownish tint.	$\rightarrow$ Sphalerite: ⎱ no distinct
	$\rightarrow$ Pitchblende: lighter.	$\rightarrow$ Pitchblende: ⎰ change.

Bireflectance/pleochroism:	Not present.
Crossed polars anisotropy:	Isotropic but some samples have a weak anomalous anisotropy.
Internal reflections:	Very dark-brown to reddish brown are frequently, although not invariably, seen.
Textures:	It forms good idiomorphic crystals and twins are not uncommon. Zonal textures are frequently noted as are lamellae in gold. Pitchblende may occur in oolitic/colloform/botryoidal/cellular and spherulitic forms.
Associated minerals:	Bismuthinite, molybdenite, galena, Ni–Co minerals, the selenides.
Other remarks:	Uraninite is crystalline, while pitchblende is massive (and often of a colloidal origin).
Density (g/cm^3):	6.5–8.5 (massive) 7.5–10.0 (crystalline)

Reflectance (%) in air:

at 546 nm	at 589 nm
13.6	13.6
Mean: 15.8; Range: 14.5–16.8	

Hardness (Mohs'):	5.5–6.0
Hardness (Talmage/polishing):	**G** (> magnetite, < pyrite).
VHN:	Mean: 765; Range: 280–839
Diagnostic features:	Very difficult to recognise as it has no diagnostic features, while pitchblende is relatively easy due to its textures.
Occurrence:	Widespread, geographically and geologically, in: granites, pegmatites, contact-metamorphic terrains as well as hydrothermal veins (especially the Co–Ni–Ag–Bi associations), placer deposits etc. Typical occurrences are: 'Colorado Plateau' deposits of the USA, Witwatersrand (S. Africa), Blind River (Ontario, Canada), Bancroft (Canada).
Etch tests:	HNO$_3$: may stain brown. FeCl$_3$: slowly tarnishes to a very dark brown-grey colour.

WOLFRAMITE

(Fe,Mn)WO$_4$ 		Monoclinic
(Wolfram)

Colour:	Dark grey to brownish black.
Streak:	Black to very dark grey.
Lustre:	Submetallic.

Colour

In air:	In oil:
Generally: grey to greyish white.	Generally: grey with brownish or yellowish tint.
→ Sphalerite: no change.	→ Sphalerite: darker.
→ Magnetite: slightly darker.	→ Magnetite: no change.
→ Cassiterite: lighter.	→ Cassiterite: no change.
→ Chalcopyrite: brownish grey.	→ Chalcopyrite: no change.

Bireflectance/pleochroism: Weak to distinct (especially at grain boundaries). It is best observed parallel to 'c' when grey to brownish grey colours are noted.

Crossed polars anisotropy: Weak to distinct. Unusual in that it is better seen in air than in oil. Yellow to dark grey and/or violet or green tints are noted. Oblique extinction.

Internal reflections: Present. Deep-red and better seen in oil.

Textures: Tabular idiomorphic crystals are common as are growth zones – with scheelite – when it is called Reinite. Twinning noted and coarse-grained examples are frequent.

Associated minerals: Scheelite, arsenopyrite, stannite, gold, bismuthinite, molybdenite.

Other remarks: None.

Density (g/cm^3): 7.2–7.5

Reflectance (%) in air:

at 546 nm	at 589 nm
15.0–16.2	14.7–15.9
Mean: 17.1; range: 15.0–20.2	

Hardness (Mohs'): 4.0–4.5

Hardness (Talmage/polishing): E (≪ cassiterite, pyrite and arsenopyrite; > magnetite and scheelite).

VHN: Mean: 436; range: 285–657.

Diagnostic features: Its lamellar form, internal reflections and low reflectance all help to identify the mineral.

Occurrence: Frequent and characteristic in pneumatolytic deposits (sometimes with quartz and cassiterite). It occurs in hydrothermal vein associations with stannite, arsenopyrite, sphalerite and chalcopyrite.
Typical occurrences are: Cornwall (UK), Boulder (Colorado, USA), and various localities in Bolivia.

Etch tests: Negative to all reagents.

APPENDIX 1 COMMON MINERAL ASSEMBLAGES OR ASSOCIATIONS

The ore minerals are unevenly distributed and occur in characteristic 'relationships' with respect to their mineralogy, textures and geological setting. These associations or assemblages may therefore assist in anticipating the minerals which may be present; it is important to note, however, that unusual or unexpected minerals may also occur.

Some ore deposits can be classified under more than one heading. For example, the Ok Tedi prospect or the Mount Fublian orebody in Papua New Guinea may be considered to be either a porphyry copper deposit or a disseminated gold deposit. Most so-called 'lead–zinc–copper' deposits may contain silver but may not contain equal amounts of lead, zinc or copper. The following subdivisions are therefore not strict in their nomenclature but are given as a guide to the ore minerals and associated minerals which may (or may not) be present. They also do not imply a genetic relationship, although a similar origin is possible. Additional information with respect to this section may be obtained by reference to: Uytenbogaardt and Burke (1973), Stanton (1972), Ramdohr (1980), Craig and Vaughan (1981) and Picot and Johan (1982).

ARSENIC–ANTIMONY OR MERCURY BASE METALS VEINS

This is a very broad group which includes the following subdivisions:

COPPER–ZINC–ARSENIC

Mineralogy
Major: pyrite, chalcopyrite, bornite, tennantite, sphalerite, enargite.
Minor: covellite, hematite, magnetite, chalcocite, digenite, galena, molybdenite.
Gangue or secondary: quartz, siderite, calcite, barite, rhodochrosite.

Examples
Bor, Yugoslavia.
Butte, Montana, USA.
Magma mine, Arizona, USA.
Tsumeb, S. Africa.

LEAD–ZINC–SILVER–ANTIMONY–ARSENIC–COPPER –SELENIUM

Mineralogy
Major: galena, sphalerite, tetrahedrite, chalcopyrite.
Minor: pyrite, pyrrhotite, arsenopyrite, magnetite.
Gangue or secondary: quartz, siderite, dolomite, barite.

Examples
Freiberg, E. Germany.
Coeur d'Alene, Idaho, USA.
Falum, Sweden.
Kapnik, Czechoslovakia.
Kuusamo, Finland.

ARSENIC–ANTIMONY–MERCURY

Mineralogy
Major: realgar, orpiment, pyrite, stibnite, cinnabar.
Minor: chalcopyrite, arsenopyrite, gold, marcasite.
Gangue or secondary: quartz, calcite, sericite, chlorite.

Examples
Allchar and Binn, Switzerland.
Elbrus, Caucasus, USSR.
Getchell, Nevada, USA.
Příbram, Czechoslovakia.
Uj-Moldava, Roumania.

MERCURY SULPHIDE ORES

Mineralogy
Major: cinnabar, metallic mercury, pyrite, marcasite.
Minor: stibnite, sphalerite, pyrrhotite.
Gangue or secondary: quartz, chalcedony, barite, dolomite.

Occurrences
The deposits often occur as strata-bound ores, and precipitation into a sedimentary basin from submarine hot-springs may be evoked. They may also be epigenetic in origin and localised along major faults. An affinity to carbonates, sandstones and shales as well as volcano–sedimentary host rocks in which they predominantly infill breccias is a common occurrence.

Examples
Almaden, Spain.
Amedee Hot Springs, New Iria, New Almaden, Sulphur Bank in California; Steamboat Springs and Cordero in Nevada; USA.
Amiata, Italy.
Huancavalica, Peru.
Idira and Trsče, Yugoslavia.
Palawan, Philippines.

CHROMITE ORES OR CHROMIUM–NICKEL–PLATINOID ± COPPER

Mineralogy
Major: chromite, pentlandite, pyrrhotite, Pt group metals.
Minor: chalcopyrite, bornite, rutile.
Gangue or secondary: Pt sulphides, arsenides, antimonides and the mafic and ultramafic minerals.

Occurrences
These ores may occur in either layered basic magmatic 'sedimentary strata' with a mean gabbroic composition (in tectonically-stable environments) or in orogenic belts associated with peridotites or serpentinised peridotites. The latter group are often referred to as 'podiform' or 'Alpine-type' chromite deposits.

Examples
Rustenburg etc. in the Bushveld Complex, Transvaal, S. Africa.
Insizwa range, Cape Province, S. Africa.
Stillwater complex, Montana, USA.
Great Dyke, Zimbabwe, S. Africa.
Guleman–Soridag district in Turkey.
Molodezhnyy in W. Kazakhstan, USSR.
Masinloc, Philippines.
Sagua de Tanamo, Cuba.
Skoumtsa and Eretria, Greece.
Kalia etc in Orrisa, India.
Marias Kiki, New Caledonia.
Kakopetria, Cyprus.
Andriamena, Madagascar.
Kemi, Finland.
Also in Pakistan, Yugoslavia, Brazil, USA, Canada and Australia.

GOLD VEINS AND RELATED MINERALISATION

Mineralogy
Major: native gold and gold tellurides.
Minor: pyrite, marcasite, arsenopyrite, pyrrhotite, galena, sphalerite, chalcopyrite, stibnite, tetrahedrite–tennantite, realgar, silver and scheelite.
Gangue or secondary: quartz, calcite, siderite, ankerite, dolomite, fluorite, barite, graphite, felspars, tourmaline, epidote, carbon and carbonaceous material.

Occurrences
In one of the widest ranging environments which may include: volcanic breccias, skarns, hydrothermal veins, disseminated and massive sulphides, veins in metamorphosed volcanics, deformed terrains and volcanic rocks.

Examples

Amur, Yakut and the Trans-Baikal region (e.g. Beli mine); Khakandzha and Karamken, E. Siberia; USSR.

Bendigo and Barrarat, Victoria; Kalgoorlie's 'Golden Mile', W. Australia; Mt Morgan, Mt Isa and Charters Towers, Queensland; Cobar, NSW, Australia.

Ok Tedi, Papua New Guinea.

Otago, Hauraki and Thames, New Zealand.

Kolar, Mysore, India.

California, Nevada, Colorado, New Mexico, USA.

Homestake Mine, S. Dakota, USA.

Hemlo camp, Kirkland Lake, Porcupine district in Ontario, Canada.

Yellowknife, Con, Giant, Salmita, Lupin, Cullaton Lake, Ptarmigan, Bullmoose and Shear Lake in the Northwest Territories, Canada.

Nagyag, Roumania.

Guanajuato, Mexico.

Tavua, Fiji.

Chinkuashih, Taiwan.

Lepanto, Philippines.

Raub Australian Mine, Malaysia.

GOLD–URANIUM AND URANIUM–GOLD

Mineralogy

Major: gold, pyrite, silver, uraninite ('thucholite') and pitchblende.

Minor: marcasite, pyrrhotite, sphalerite, galena, molybdenite, arsenides and sulpharsenides.

Gangue or secondary: quartz, sericite, chlorite, chromite, zircon, ilmenite, magnetite, rutile, hematite and calcite.

Occurrences

As disseminated beds or lenses of coarse conglomerates in arenaceous sequences. The quartz pebbles may be of a fluviatile or shallow-water deltaic origin. These deposits are considered to be placer, palaeo-placer, modified placer or hydrothermal in origin.

Examples

Witwatersrand (Au + U), S. Africa.

Hemlo Camp, Agnew Lake, Elliot Lake or Blind River deposits (U ± Au) in Ontario, Canada.

Jacobina (Au), Brazil.

Numerous deposits in the Yukon, Canada.

Athabasca Basin, N. Saskatchewan, Canada, i.e. the Rabbit Lake, Key Lake and Beaverlodge mining districts.

Lena-Amur, USSR.

Deposits in Columbia, Bolivia, Peru and New Zealand.

GOLD–SILVER TELLURIDES

Mineralogy
Major: gold, silver tellurides, silver and tellurium.
Minor: selenides, base metal sulphides and sulphosalts.
Gangue or secondary: pyrite, quartz, calcite, barite, anhydrite and gypsum.

Examples
Acupan, Philippines.
Cripple Creek, Colorado, USA.
Kirkland Lake and Porcupine districts, Ontario, Canada.
Kalgoorlie, W. Australia.
Noranda, Quebec, Canada.
Tavua, Fiji.
Teine Mine, Hokkaido, Japan.

GOSSANS OR 'IRON CAPS'

Mineralogy
Major: limonite, goethite, hematite and lepidocrocite.
Minor: manganese oxides and residual base metal sulphides.
Gangue or secondary: any constituent of the country rock or the ore deposit.

Occurrences
Developed on most iron sulphide-bearing deposits as a result of surface or near-surface weathering or oxidation.

Examples
World-wide at surficial exposures of sulphide-containing deposits.

IRON–TITANIUM OXIDES

Mineralogy
Major: magnetite, ilmenite and hematite.
Minor: pyrite, chalcopyrite, pyrrhotite and rare Fe and Ti minerals.
Gangue or secondary: hematite, rutile, maghemite, actinolite, diopside or calcite.

Occurrences
These deposits occur as layers or massive thick sheets or lenses in anorthosites, gabbros and norites. They may also occur as accessory minerals in acid, intermediate and basic igneous rocks.

Examples
As well as the Bushveld, Stillwater and Duluth complexes, this assemblage/association occurs at:

Allard Lake, Quebec; and Wilson Lake, Labrador, Canada.
Egersund, Sogndal, Baugsto and Larvik, Norway.
Kiruna (Kiirunavaara and Luosavaara deposits), in the Arvidsjaur complex, Sweden.
Tahawus, New York State; Pea Ridge, S.E. Missouri, USA.

IRON–NICKEL–COPPER SULPHIDES ± PLATINOIDS

Mineralogy
Major: pyrrhotite, pentlandite, pyrite, magnetite and chalcopyrite.
Minor: cubanite, platinoids (platinum, palladium, rhodium, ruthenium, osmium and iridium), chromite, ilmenite and molybdenite.
Gangue or secondary: millerite and the mafic and ultramafic minerals.

Occurrences
These deposits occur as massive or disseminated ores in close association with mafic and ultramafic intrusive and extrusive complexes. They may also occur in metamorphosed mafic and ultramafic rocks.

Examples
Bushveld complex (Merensky reef), Transvaal, S. Africa.
Insizwa complex, Cape Province, S. Africa.
Sudbury Basin, Texmont and Gordon Lake, Ontario, Canada.
Thompson Mine (Moak Lake–Setting Lake) and Lynn Lake, Manitoba, Canada.
Ungava district, Marbridge and Lorraine areas of Quebec, Canada.
Hanna Mine in Oregon and Cerro Matosa, Columbia; Lick Fort, Virginia, USA.
Kambalda area, W. Australia (Windarra, Wannaway and Nepean Mines).
Pechenga and Monchegorsk (Kola) and Noril'sk, Talnakhski and Octiabrski in Siberia, USSR.
Cerro Matosa, Columbia.
Trojan, Shangani, Madziwa and Epoch Mines, Zimbabwe.

COPPER ± ZINC–IRON–SILVER
ZINC ± COPPER–IRON
ZINC–LEAD ± COPPER–IRON

Mineralogy
Major: pyrite, sphalerite, chalcopyrite ± pyrrhotite or galena.
Minor: bornite, tetrahedrite, electrum, marcasite, cubanite, copper–lead–bismuth–silver sulphosalts, cassiterite, arsenopyrite and bismuth.
Gangue or secondary: numerous other minor constituents including quartz, cryptocrystalline silica, barite, hematite, chlorite, anhydrite and argentite.

Occurrences

These ores occur as massive bodies or disseminated stratiform sulphides in a volcano-sedimentary environment, e.g. ophiolite complexes (Cyprus-type), felsic tuffs, lavas as well as sub-sea intrusions (Kuroko-type). Mudstones and shales with a minor volcanic component are also included in this assemblage (Besshi-type).

Examples

Cyprus-type (iron rich)
Troodos complex, Cyprus.
Betts Cove, Newfoundland, Canada.
Lorraine orebody, Balabac island, Borneo.
Barlo Mine, West Luzon, Philippines.
Ergani-Maden, Turkey.

Kuroko-type (copper–zinc rich)
Kuroko and the Hokuroko district (mines at Hanaska, Kosaka and Shakanai), Japan.
Red Sea and East Pacific Rise deposits.
Iberian Pyrite Belt of Spain and Portugal, especially the Rio Tinto area.
Avoca, Eire.
Parys Mountain, Anglesey, UK.
Vanua Levu, Fiji.
Buchans, Newfoundland, Canada.

Besshi-type
Besshi and Shikoku, Japan.

BANDED IRON FORMATIONS

Mineralogy

Major: hematite, magnetite $\pm$ iron carbonates and silicates.
Minor: pyrite.
Gangue or secondary: chert and jasper.

Occurrences

Laterally and vertically extensive Precambrian (predominantly) bedded, alternating chert–iron stratified sequences. The primary setting may have been marine, estuarine or lacustrine, although most deposits have undergone considerable secondary near-surface leaching/concentration which has up-graded the ores on the palaeo-surface.

Examples

Lake Superior region (Minnesota–Winsconsin–Michigan–Ontario of the USA and Canada). Specific areas include: Iron River, Crystal Falls, Marquette, Gogebic, Menominee, Mesabi, Gunflint and Biwabik.
Labrador province of Canada: Ungava Bay and Wabush Lake deposits.
Pilbara and Hamersley ore provinces of W. Australia which include deposits at:

Mt Tom Price, Mt Newman, Mt Whaleback, Mt Goldsworthy, Pilbara and Paraburdoo.
Middleback ranges of S. Australia.
Rana Gruber Mine, Norway.
Krivoi Rog series of the Ukraine; the Verkhtsevskovo region, USSR.
The Quadrilatero Ferrifero area in Minas Gerais (Brazil).
Nimba Mine, Liberia.
Cerro Bolivar, Venezuela.
Other deposits are found in India and S. Africa (Witwatersrand).

BOG AND BLACKBAND IRON ORES

Mineralogy
Major: goethite, limonite, siderite and carbonates.
Minor: manganese oxides and trace elements.
Gangue or secondary: clays and other detritus.

Occurrences
These ores occur in a lake, peat or marsh environment in a temperate or recently glaciated area. Volcanic lake ores and the 'blackband ironstones' are included in this category.

Examples
Tundra areas of Scandinavia and Canada.
Temperate coastal areas of the USA (Ohio) and Canada.
Volcanic province of the Kurile Islands, Japan.
Carboniferous and Permian sequences of the eastern USA and the UK.
Prestwich Mine, Natal, S. Africa.

SEDIMENTARY IRONSTONES

Mineralogy
Major: chamosite, limonite, goethite and siderite.
Minor: magnetite, pyrite and collophane.
Gangue or secondary: glauconite.

Occurrences
The sedimentary ironstones were deposited in shallow-water marine sequences and as such are associated with limestones, siltstones, shales, sandstones and (possibly) volcanics. The type examples are the Clinton or 'minette-type' of the USA.

Examples
Ordovician
Anglesey, UK.
Wabana, Newfoundland, Canada.

Devonian
Gila County, Arizona, USA.
Lahn-Dill region (Koenigszug and Constanze Mines) of W. Germany.

Silurian:
Clinton, Kirkland and Westmoreland (New York), USA.

Carboniferous
Numerous deposits in the UK.

Jurassic
Alsace-Lorraine, France, as well as in Germany, Belgium and Luxemburg.

LEAD–ZINC DEPOSITS IN SEDIMENTS (ESPECIALLY CARBONATES)

Mineralogy
Major: galena, sphalerite ± barite and fluorite.
Minor: pyrite, marcasite, chalcopyrite, bornite, covellite, enargite, hematite, cuprite, limonite, smithsonite, cerussite, malachite and anglesite.
Gangue or secondary: calcite, dolomite, siderite, ankerite, aragonite, quartz and chert.

Occurrences
Lead–zinc deposits can occur in most sediments (e.g. sandstones, shales and conglomerates), the predominant host rocks being carbonates, especially dolomites, limestones and magnesian limestones. The ore deposits can be either stratiform, vein, fissure, joint, solution collapse breccias (related to karst or palaeoaquifer development) or fault infillings. They have been subdivided into: 'Mississippi Valley-type' and 'Alpine-type' depending on their depositional environment.

Examples
The 'Tri-State' field of Missouri–Kansas–Oklahoma; the Old and New Lead Belts (= Verburnum Trend) of S.E. Missouri; S. Appalachians and E. Tennessee deposits (Mascot–Jefferson City, Copper Ridge and Sweetwater districts); S. Illinois (Cave-in-Rock area)–Kentucky areas of the USA.
The Pennine deposits of the UK.
The Silesia–Cracow Pb–Zn area of Poland.
Daniels Harbour, Newfoundland, Canada.
Bleiberg Pb–Zn belt of Switzerland–Austria.
Pine Point and Polaris deposits of the Northwest Territories, Canada.
Siberian platform of the USSR (e.g. Kazakhstan).
Laisvall, Norway.
Trento Valley, Italy.
Navan, Co. Meath and Skerries in Eire.
Burs, Austria.
Les Malines, France.
Alpujarrides of Spain.

Touissit and Bou Beker of Morocco.
El Abed and Deglen of Algeria.
Duna, Central Alborz, Iran.
Breithorn, E. Greenland.

BASE METAL SULPHIDES IN SEDIMENTS AND VOLCANICS

(in which copper and lead–zinc predominate)

Mineralogy
Major: pyrite, chalcopyrite, galena, sphalerite, pyrrhotite, bornite, chalcocite, copper and other sulphides and sulpharsenides.
Minor: arsenopyrite, tetrahedrite, bismuth, bismuthinite, argentite, niccolite, covellite and molybdenite.
Gangue or secondary: barite, fluorite and other carbonates.

Occurrences
These ores occur as disseminated or massive stratiform deposits which are laterally extensive and conformable with the sedimentary sequences. The host rocks may be black shales, dolomites, arenites, arkoses and quartzites which may or may not have undergone folding and metamorphism. Occasionally, volcanic deposits may predominate in the stratigraphical sequence. Synsedimentary/syndiagenetic depositional environments have been postulated.

Examples
The copper belt of N. Zambia and S. Zaire (Luanshya, Nkana, Chambishi, Chingula, Nchanga and Mufulira Mines).
The Kupferschiefer or Marl Slate of northern Europe.
Rammelsberg, Germany.
Coeur d'Alene, Idaho; White Pine Mine, Michigan, USA.
Selwyn Basin in the Yukon; Sullivan Mine, British Columbia, Millenbach, Quebec; Canada.
Broken Hill, NSW; Mt Isa and Hilton, Queensland and the McArthur River or HYC deposit in the Northern Territories, Australia.
Navan, Co. Meath, Eire.
Raibl, Friuli, Italy.

LEAD–ZINC–COPPER, COPPER–LEAD–ZINC–SILVER

Mineralogy
Major: pyrite, sphalerite, galena, chalcopyrite and tetrahedrite.
Minor: bornite, chalcocite, enargite, gold, argentite, proustite, pyrargyrite, hematite, pyrrhotite and Pb–Bi–Sb sulphosalts.
Gangue or secondary: azurite, malachite, cerussite, anglesite, goethite, smithsonite, covellite and silver.

Occurrences

Hydrothermal vein and replacement deposits (usually in limestones) are the most frequent occurrences; however, they may be associated with acid and intermediate intrusive rocks, develop skarn aureoles and illustrate mild folding and metamorphism. As this is a large group, it encompasses deposits which are included in other subdivisions.

Examples

Bingham, Tintic, Park City in Utah; Creede, Gilman, Leadville in Colorado; Tonopath, Eureka, Comstock Lode in Nevada; Balmat-Edwards, Pierrepont in New York State; Coeur d'Alene district and especially Sunshine and Galena Mines in Idaho; Sterling and Franklin Furnace in New Jersey, USA.

Sullivan, Bluebell, Myra Creek, Goldstream and Lynx Mines in British Columbia; Buchans in Newfoundland; Cobalt, Ontario; Polaris and Nonisivik Mines in the Northwest Territories, and Faro in the Yukon of Canada.

Fresnillo, Zacatecas, Potosi and Chihuahua Mines in Mexico.

Casapalca in Peru.

Potosi district in Bolivia.

E. Transbaikalia in the USSR.

Raibl, Trentino and Trieste Mines in Italy.

The Arbus district in Sardinia.

Los Guindos, La Roas and Arrayanes lode in Spain.

Trepča district and Stantrg Mine in Yugoslavia.

Kupferschiefer–Marl Slate of Europe, especially the Mansfeld (Harz) district of Germany.

Tsumeb, Kombat, Asis West and Matchless Mines in Namibia, S. Africa.

Mt Isa, Queensland; Broken Hill, Cobar, Elura and Woodlawn Mines in New South Wales; Rosebury and Captain's Flat in Tasmania (Australia).

MANGANESE DEPOSITS ASSOCIATED WITH VOLCANIC AND SEDIMENTARY SEQUENCES

Mineralogy

Major: psilomelane, pyrolusite, hematite, siderite and magnetite.

Minor: pyrite, arsenopyrite, chalcopyrite, galena, sphalerite, calcite and tetrahedrite.

Gangue or secondary: quartz, barite and chalcedony.

Occurrences

This assemblage occurs as concordant lenses in volcanic and sedimentary sequences.

Examples

Chario Kendondo and Quinto Mines, Cuba.

Skouriostissa, Mathiati and Mousoulos Mines, Cyprus.

Nsuta Mine, Ghana.

Kokuriki, Nikura and Hokkaido areas, Japan (e.g. Noda-Tamagawa Mine).

Semail nappe area, Oman.
Lake Superior iron province, USA and Canada.
Palaeozoic pyroclastic sequence of W. Australia.
Karadzhal and C. Kazakhstan areas of the USSR.
Also in E. Australia, Indonesia, West Indies and the Philippines.

MANGANESE DEPOSITS ASSOCIATED WITH LIMESTONES AND DOLOMITES

Mineralogy
Major: pyrolusite and psilomelane.
Minor: galena, pyrite, pyrrhotite, calcite, rhodochrosite, anthracite–graphite.
Gangue or secondary: silica and the heavy metals.

Occurrences
The deposits are associated with limestone and/or dolomite sequences on a stable platform. They are often interlayered with gypsum and have red terruginous sequences beneath the deposits. Surface erosional features are characteristic of these ores. They are classified as 'Moroccan-type' rather than the 'Appalachian or Usinsk-type' which were formed in geosynclinal zones and do not have terruginous or erosional surfaces. Lavas and bedded iron formations may be interbedded with the sediments. Supergene-enriched derivatives of the primary ore are mined. Some deposits are metamorphosed.

Examples
Imini, Morocco.
Kalahari Manganese Field, S. Africa (e.g. Mamaturan and Lohathla Mines).
Madhya Pradesh and Maharashtra areas of India.
Appalachian area, USA.
Usinsk deposits of S.W. Siberia, USSR.

MANGANESE DEPOSITS ASSOCIATED WITH GLAUCONITIC CLAYS AND ORTHOQUARTZITES

Mineralogy
Major: pyrolusite, psilomelane, manganite, manganocalcite.
Minor: rhodochrosite.
Gangue or secondary: clay minerals, silica and glauconite.

Occurrences
The ores normally occur as thin lens-like deposits associated with sedimentary strata on a stable platform. The sediments may be estuarine, shallow marine, limestones, silts or marls.

Examples
Chiatura, Nikopol', Labinsk and Bol'she-Tokmaksk in the USSR.

Groote Eylandt, N. Australia.
Timna, Israel.
Also in Turkey and Bulgaria.

MANGANESE NODULES

Mineralogy
Major:　manganese and iron oxides.
Minor:　goethite, quartz, clay minerals, felspars and zeolites.

Occurrences
The nodules are found on the ocean floors. They are normally spherical with diameters of between 1 and 50 cm. They contain appreciable amounts of cobalt, nickel and copper. Often associated with manganese-rich sediments (oozes), they appear to have a relationship with submarine volcanism. Nodules are reported from the Pacific Ocean as well as the Baltic, White, Barent and Kara Seas.

MERCURY–ANTIMONY ± TUNGSTEN ± GOLD ORES

Mineralogy
Major:　cinnabar, realgar, stibnite and scheelite.
Minor:　pyrite, marcasite, gold, chalcopyrite, sphalerite, galena and bornite.
Gangue or secondary:　silica, chalcedony, calcite, silicates and carbonate minerals.

Occurrences
The ores occur as veins, stringers or cements to the host rocks which may be sedimentary sequences with or without volcanic horizons (tuffs and sinters). They are also encountered around hot springs.

Examples
Almaden and Badajoz, Spain.
Idrija and Idria (or Idrijia), Yugoslavia.
Monte Amiate Mine, Italy.
Hepeh and W. Hunan areas of China.
Bau region of Sarawak, Borneo.
New Almaden district, California, USA.
Palawan island, Philippines.

TIN–TUNGSTEN–BISMUTH ± SILVER–GOLD

Mineralogy
Major:　cassiterite, stannite, scheelite, arsenopyrite, wolframite, bismuthinite, pyrite, marcasite, pyrrhotite, gold, silver and bismuth.

Minor: chalcopyrite, sphalerite, tetrahedrite, pyrargyrite, galena, molybdenite, ilmenite and magnetite.

Gangue or secondary: quartz, tourmaline, apatite, calcite, barite and fluorite.

Occurrences

A wide range of occurrences are exhibited by this assemblage; however, the most common one is as hydrothermal vein filling (with a complex paragenesis). The deposits are often associated with granites. In some tungsten–bismuth deposits, tin may be absent.

Examples

Tin belt of Malaysia, Thailand and Indonesia.

The tin belt of Bolivia (Potosi, Oruro and Llallangua districts) as well as Tasna, Chacaltaya, Colcha, Huanuni, Sayaquira and Mina Fobulosa Mines and areas).

Pirquitas, Argentina.

Jukani et Hualgoyoi and Pasto Buena, Peru.

Hamme district, N. Carolina, USA.

S. Kiangsi, N. Kwang Tung, S.E. Hunan and E. Kwantung areas of China.

Panasqueira, Portugal.

Bustarviejo, Spain.

Cornwall, UK.

Erzgebirge, E. Germany.

Krnsne Hory, Czechoslovakia.

Namaqualand, S.W. Africa.

Bindal, Norway.

Sangdong Mine, S. Korea.

King Island, Renison-Bell and Cleveland, Mt Bischoff,

Tasmania, Australia.

COPPER–MOLYBDENUM ASSOCIATED WITH PORPHYRY INTRUSIVE IGNEOUS ROCKS (i.e. the 'Porphyry Copper–Molybdenum' deposits)

Mineralogy

Major: pyrite, chalcopyrite, molybdenite and bornite.

Minor: magnetite, hematite, ilmenite, enargite, cubanite, cassiterite, gold and stannite.

Gangue or secondary: covellite, copper and rutile.

Occurrences

The porphyry copper–molybdenum deposits occur as veinlets or disseminated grains in, or adjacent to, porphyritic quartz diorites to quartz monzonites. The host rock develops a characteristic alteration assemblage, i.e. siliceous to potassic to phyllic to argillic to propylitic zonation.

Examples

Butte, Montana; Bingham, Utah; Ajo, Bisbee, Ray, Morenci, Safford, San

Manuel–Kalamazoo and Red Mountain, Arizona; Climax (Mo), Henderson (Mo) and Urad (Mo), Colorado; Questa and Santa Rita, New Mexico, USA.
El Salvador, Chuquicamata, El Tentiente, Braden and Andina, Chile.
Cerro de Pasco, Peru.
Cananea and La Caridad, Mexico.
Bethlehem and Endako (Mo), British Columbia, Canada.
Cerro Colorado, Panama.
Sar Cheshmeh, Iran.
Rio Tinto, Spain.
Kal'makyr deposit in Almalyk region of the USSR.
Panguna on Bougainville; Mt Fublian orebody or the Ok Tedi prospect, Papua New Guinea.
Atlas on Cebu, Sipalay on Negros Occidental, Philippines.
Copper Hill and Cobar–Mt Hope district of NSW, Australia.

SILVER–COBALT–NICKEL ± ARSENIC–COPPER

Mineralogy
Major: silver, bismuth, niccolite, löllingite, cobaltite, arsenopyrite, pyrite, marcasite, chalcopyrite, galena, pyrrhotite, sphalerite and uraninite.
Minor: tetrahedrite, Fe, Co and Ni in combination with As, S and Sb minerals (i.e. sulphides, sulphosalts).
Gangue or secondary: quartz, calcite and carbonates.

Occurrences
This assemblage is often referred to as the 'Cobalt-type'. The ores occur in fissure fillings, faults and joint planes. The host rocks may be quartzites, greywackes, conglomerates, schists, granites, slates and diabases.

Examples
Cobalt – Gowganda area of Ontario, the Great Bear Lake deposit of the Northwest Territories, Canada.
Jáckymov and Cĕrny Dul areas of Czechoslovakia.
Chalanches, France.
Erzgebirge, E. Germany.
Kongsberg, Norway.
Batopilas and Sabinal districts of Mexico.
Sarrabus, Sardinia.
Traver Mine, Sweden.
Turtmannstal, Switzerland.

URANIUM–VANADIUM–COPPER ORES

(may occur as U or U–Cu, U–V–Cu but not Cu–V)

Mineralogy
Major: uraninite or pitchblende, vanadium-bearing mixed-layer clay minerals, chalcocite, bornite, chalcopyrite, covellite and copper.
Minor: pyrite, galena, sphalerite, molybdenite, gold and silver.
Gangue or secondary: numerous oxides, hydrated oxides carbonates and sulphates.

Occurrences
Irregular masses which infill faults, shear zones, breccias, pores and veins and replace organic material are typical occurrences. The deposits may also be associated with major unconformities related to continental weathering as well as conglomerates, sandstones and siltstones. They are frequently found with calcrete and red beds.

Examples
Colorado Plateau deposits of Colorado, Arizona, Utah, New Mexico and Wyoming, USA.
Darwin area of Australia; typical deposits are: Alligator River, Rum Jungle, Jabiluka, Ranger One, Koongarra and Nabarlek (U-rich).
Athabasca basin (Rabbit Lake and Key Lake) of N. Saskatchewan, Canada.
Corocoro (Cu-rich), Bolivia.
Udokan (Cu-rich), Siberia, USSR.
Yeelirrie, W. Australia.
Langer Heinrich, Namibia, S.W. Africa.
Dusa Mareb, Somalia, E. Africa.
Mary Kathleen, Queensland, Australia.

URANIUM ASSOCIATED WITH GRANITOIDS ± TIN–TUNGSTEN

Mineralogy
Major: uraninite, pitchblende, tin and tungsten minerals.
Minor: pyrite, chalcopyrite, bornite, molybdenite, arsenopyrite, magnetite, hematite and fluorite.
Gangue or secondary: marcasite, quartz and iron oxides.

Occurrences
As veins, fissures or disseminated ores in S-type granites or in metamorphosed aureoles as well as in high-grade metamorphic terrains.

Examples
Hercynian granite chain of Europe (Spain to the Erzgebirge) with particular reference to the deposits in the Massif Central of France.

Erzgebirge (E. Germany).

Příbram and Jáchymov, Czechoslovakia.

Bor, Yugoslavia.

Rossing, Namibia, S.W. Africa.

Midnite and Spokane Mountain Mines, Washington, USA (excludes Mary Kathleen in Queensland, Australia).

ORES IN METAMORPHIC TERRAINES

Mineralogy

Major: pyrite, pyrrhotite, sphalerite, chalcopyrite, galena and tetrahedrite.

Minor: cubanite, marcasite, ilmenite, magnetite and arsenopyrite.

Gangue or secondary: any of the metamorphic minerals or such minerals formed by the interaction of the ores and the country rocks.

This group may be subdivided into four metamorphic settings:

IN REGIONAL METAMORPHIC ROCKS AND ESPECIALLY VOLCANIC SEQUENCES

Examples

Ducktown, Tennessee; Ore Knob, N. Carolina; Great Gossan Lake, Virginia; USA.

Flin Flon, Manitoba; Buchans, Newfoundland; Sullivan, British Columbia and Bathurst, New Brunswick; Canada.

Mt Isa, Queensland and Broken Hill, NSW Australia.

Skellefte District, Sweden.

Sulitjelma and Røros, Norway.

Rammelsberg, W. Germany.

Minas Gerais area, Brazil.

DUE TO METASOMATIC METAMORPHISM

Examples

Mt Isa, Queensland, Australia.

Broken Hill, NSW, Australia.

Sullivan Mine, British Columbia, Canada.

Copper belt of Zambia and Zaire.

DUE TO DISLOCATION METAMORPHISM

Examples

Broken Hill, NSW, Australia.

Coeur d'Alene, Idaho, USA.

SKARNS OR TACTITES

These deposits are coarse-grained ores formed due to contact-metamorphic haloes/aureoles adjacent to intrusions in carbonates or Al and Si-rich rocks. They may develop massive deposits or disseminated grains and veinlets. Reaction skarns have a restricted width and are predominantly Mn silicates and carbonates. Replacement skarns give rise to large areas of silicate-rich rocks and the replacement of carbonate host rocks. They frequently contain the ores of Fe, Cu, Zn, W and Mo.

Examples

Christmas and Morenci-Metcalf Mines (Cu), Arizona; Eagle Mountain (Fe), Darwin (Pb, Zn and Ag), Bishop (W, Mo and Cu) Mines in California; Hanover (Pb and Zn) and Magdalena (Pb and Zn), New Mexico; Cotopaxi (Cu, Pb and Zn), Colorado; Buena Vista (Fe), Ely (Cu), Nevada; Cornwall and Morgantown (Fe), Pennsylvania, Iron Springs (Fe), Bingham (Cu), Utah; Lost Creek (W), Montana; USA.

Murdockville (Cu), Gaspe Copper (Cu), Quebec; Cantung (W), Northwest Territories; Canada.

King Island and Renison-Bell (W), Tasmania, Australia.

Cananea (Pb and Zn), Mexico.

Kinta Valley tin and tungsten skarns of Malaysia.

S.E. Hunan (W), China.

DIAGRAM OF OPTICAL DETERMINATIONS OF THE ORE MINERALS ON THE BASIS OF REFLECTANCE VALUES (FOR 589 nm) AND MICROHARDNESS (VHN)

Reproduced with modifications and the publisher's permission from Tarkian (1974).

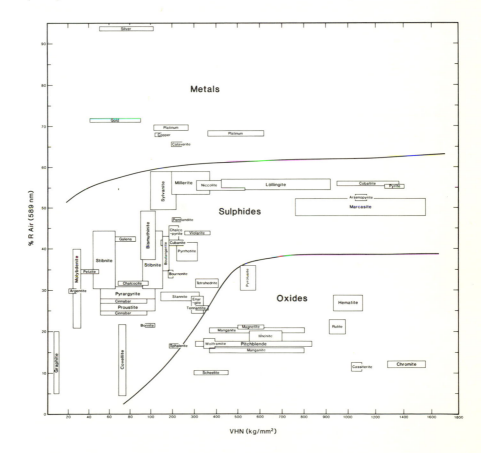

REFLECTED LIGHT OPTICS

Introduction

The theory of reflected light optics for anisotropic media is complex and cannot be explained briefly and at the same time usefully. Below, those principal results of theory are stated (without derivation) which are made use of in the present text. For a more detailed treatment, the reader is referred to such texts as Cameron (1961), Galopin and Henry (1972) and Craig and Vaughan (1981).

REFLECTION OF PLANE POLARISED LIGHT

MONOCHROMATIC LIGHT

Cubic minerals and glasses

For these substances, all cross-sections are optically equivalent and isotropic. The percentage of reflectance ($R\%$) has only one value whatever the vibration direction of the incident light, and this value is characteristic of the mineral or glass.

Tetragonal, hexagonal and trigonal minerals (uniaxial)

Basal sections of these minerals give isotropic surfaces so that the reflectance R_o corresponding to the ordinary vibration direction has one value whatever the vibration direction of the incident light.

Prism sections give anisotropic surfaces the reflectances which are R_e perpendicular to the unique axis and R_e parallel to this direction. If $R_o < R_e$ the mineral is uniaxial positive and if $R_o > R_e$ it is negative. The difference $R_e - R_o$ is the **bireflectance**.

For other surfaces of uniaxial minerals, two reflectances are observed: one is R_o (vibration perpendicular to the plane containing the optic axis and the microscopic axis) and the other (vibration perpendicular to R_o) has a reflectance percentage lying between R_o and R_e.

Orthorhombic minerals

For these minerals, each of the three principal orthorhombic directions (x, y, z) has a characteristic reflectance (R_x, R_y, R_z) for vibrations parallel to it. Thus {100}, {010} and {001} sections will each give three reflectances R_yR_z, R_xR_z

and R_xR_y. For orthorhombic minerals it is usual to report the minimum and maximum R values, and to call them R_1 and R_2.

For **monoclinic and triclinic minerals** the symmetry of optical properties is lower, but again it is customary to report only the maximum and minimum reflectances R_1 and R_2.

WHITE LIGHT

Cubic minerals and glasses
When white light is used, the reflectance can vary with wavelength so that a coloured effect may result. A curve of R against wavelength (called the **spectral dispersion curve**) can be plotted.

Anisotropic minerals
For **uniaxial minerals**, R_o and R_e may be plotted against wavelengths, and if they do not vary in exactly the same manner, the bireflectance, $R_e - R_o$, will also vary with wavelength.

For **biaxial minerals**, R_1, R_2 and $(R_1 - R_2)$ may all vary with wavelength. These phenomena are described as '**reflection pleochroism**'.

REFLECTION BETWEEN CROSSED POLARS

MONOCHROMATIC LIGHT

Isotropic surfaces (cubic minerals and basal sections of uniaxial minerals)
For these surfaces, the plane of polarisation of incident light is unchanged by reflection and it is therefore cut out by the analyser in the 'crossed' position. Ideally the section appears black ('extinguished') in all orientations, but some light may be seen due to internal reflections and surface imperfections, and also faint illumination may result (especially at high magnifications) if the incident beam is not perfectly normal to the surface. The mineral remains extinguished while it is rotated between crossed polars.

Non-basal sections of uniaxial minerals, and sections parallel to planes of symmetry or perpendicular to diad axes for other minerals
These are known as symmetry sections, comprising $\{hk1\}$ planes for uniaxial, $\{0k1\}$, $\{h01\}$ and $\{hk0\}$ for orthorhombic and $\{h01\}$ for monoclinic minerals. Such sections have two principal vibration directions (e.g. z and $\perp z$ for $\{hk0\}$) and when either of these is aligned with the polariser or analyser the light is 'extinguished'. This situation will occur therefore every 90° on rotating the specimen between crossed polars. In between these extinction positions, the mineral will not appear dark. This phenomenon in reflected light may be termed '**crossed polars anisotropy**'.

General sections of orthorhombic, monoclinic and triclinic minerals
These sections do not extinguish at any position when rotated between crossed polars but go through minima of illumination every 90° of rotation.

WHITE LIGHT

Similar effects to those described above are produced, but variations in colour as well as intensity can occur. When some light comes through the analyser, it means that the plane of polarisation of the incident light has been rotated by the specimen. The amount of rotation can vary with orientation of the sample and with wavelength of the light. The effects are therefore called **'anisotropic rotation colours'**. The colours can also be changed by slightly uncrossing the polars, and in unsymmetrical sections the effects are different for clockwise and anticlockwise uncrossing. All of these rotation effects can be used diagnostically.

LIST OF ABBREVIATIONS

Å	Angstrom unit (= 10^{-8} cm)
c	velocity of light
cc or cm^3	volume
Eh	redox potential
eV	electron volt
F	test force in Newtons (N)
g/cm^3	density or specific gravity (SG), or g/ml
kV	kilovolts
mA	milliamps
n	refractive index
NA	Numerical aperture
nm	nanometre (= 10^{-6} mm) 1 nm = 10 Å
μm	micrometre (10^{-3} mm)
ppb	parts per billion (solid measure)
ppm	parts per million (solid measure)
R	reflectance ($R\%$)
RI	refractive index
$\mu g/ml$	micro grams per millilitre
VHN	Vickers Hardness Number

REFERENCES

Augustithis, S. S. (1982) *Atlas of the Sphaeroidal Textures and Structures and their genetic significance.* Theophrastus Publ. S.A., Athens.

Badham, J. P. N., Robinson, B. W. and Morion, R. D. (1972) The Geology and genesis of the Great Bear Lake Silver Deposits. *24th Int. Geol. Cong. (Montreal)*, Section 4, 541–8.

Barnes, H. L. (ed.) (1979) *Geochemistry of Hydrothermal Ore Deposits.* Wiley-Interscience, New York.

Barton, P. B. (1978) Some ore textures involving sphalerite from the Furutobe Mine, Akita Prefecture, Japan, *Mining Geol.*, **28**, 293–300.

Bastin, E. S. (1950) Interpretation of ore textures, *Geol. Soc. Am. Mem.*, **45**.

Baumann, L. (1976) *Introduction to ore deposits.* Scott. Acad. Press, Edinburgh and London.

Bowie, S. H. U. and Simpson, P. R. (1977) Microscopy: reflected light, pp. 109–65 in Zussman, J. (ed.) *Physical Methods in Determinative Mineralogy.* Academic Press, London.

Bowie, S. H. V. and Simpson, P. R. (1980) *The Bowie-Simpson System for the Microscopic Determination of Ore Minerals.* McCrone Research Associates, London. (Distributed by: Blackwell Scientific Publications.) (First Student Issue)

Brown, J. S. (ed.) (1967) Genesis of lead–zinc–barite–fluorite deposits – A Symposium, *Econ. Geol.*, Monograph No. 3.

Cameron, E. N. (1961) *Ore Microscopy.* John Wiley, New York.

Craig, J. R. and Vaughan, D.J. (1981) *Ore Microscopy and Ore Petrography.* John Wiley, New York.

Deer, W. A., Howie, R. A. and Zussman, J. (1962) *Rock-Forming Minerals, Vol. 5. Non-Silicates.* Longman, London.

Edwards, A. B. (1960) *Textures of the Ore Minerals and their Significance.* Aust. Inst. Min. Metall., Melbourne, Australia.

Evans, A. M. (1980) *An Introduction to Ore Geology.* Blackwell Scientific Publications, Oxford.

Freund, H. (ed.) (1966) *Applied Ore Microscopy – Theory and Techniques.* Macmillan Co., New York.

Galopin, R. and Henry, N. F. M. (1972) *Microscopic study of opaque minerals.* McCrone Research Associates, London.

Hagni, R. D. and Grawe, O. R. (1964) Mineral Paragenesis in the Tri-State District, Missouri, Kansas, Oklahoma. Econ. Geol, **59**, 449–57.

Henry, N. F. M. (1977) *Commission on Ore Microscopy: IMA/COM Quantitative, Data File (first issue)*, Applied Mineralogy Group. Mineralogical Society, London.

Hutchison, C. S. (1983) *Economic Deposits and their Tectonic Setting.* Macmillan Press, London.

Jensen, M. L. and Bateman, A. M. (1981) *Economic Mineral Deposits* (3rd edn). John Wiley, New York.

MacKenzie, W. S., Donaldson, C. H. and Guildford, C. (1982). *Atlas of igneous rocks and their textures.* Longman, London.

McLeod, C. R. and Chamberlain, J. A. (1968) Reflectivity and Vickers Microhardness of ore minerals, Geol. Surv. Canada, Mines & Resour., Paper 68–64 (charts and tables).

Maucher, A. and Rehwald, G. (1961 onwards) *Card Index of Ore Photomicrographs.* Umschau Verlag, Frankfurt am Main.

Oelsner, O. (1966) *Atlas of the most important ore mineral parageneses under the microscope.* Pergamon Press, Oxford.

Otto, D. A. and Rensburg, W. C. J. van (1968) A standard method for the preparation of ore specimens for microscopic examination, *S. Afr. Geol. Surv.*, 5, 113–18.

Park, C. F. and MacDiarmid, R. A. (1964) *Ore Deposits.* W. H. Freeman and Company, San Francisco and London.

Picot, P. and Johan, Z. (1982) *Atlas of Ore Minerals.* B.R.G.M. and Elsevier, Orléans and Amsterdam.

Ramdohr, P. (1980) *The Ore Minerals and their Intergrowths* (2nd edn, 2 vols). Pergamon Press, Oxford.

Ribbe, P. H. (1974) *Sulfide Mineralogy – Short Course Notes.* Mineral. Soc. of America, Vol. 1.

Rieder, M. (1969) Replacement and cockade textures, *Econ Geol.*, 64, 564–7.

Roedder, E. (1968) The noncolloidal origin of 'colloidal' textures in sphalerite ores, *Econ. Geol.*, 63, 451–71.

Rumble, D. (1976) *Oxide Minerals – Short Course Notes Vol. 3.* Mineral. Soc. of America, Washington D.C.

Schneiderhöhn, H. and Ramdohr, P. (1931) *Lehrbuch der Erzmikroskopie, Vol. II*, Berlin.

Short, M. N. (1940) Microscopic determination of the ore minerals, *US Geol. Surv. Bull.* No. 914, Washington, USA.

Shouten, C. (1962) *Determination tables for ore microscopy.* Elsevier, Amsterdam.

Spry, A. (1969) *Metamorphic Textures.* Pergamon Press, Oxford.

Stanton, R. L. (1972) *Ore Petrology.* McGraw Hill.

Sweeney, R. I. and Kaplan, I. R. (1973) Pyrite framboidal formation, *Econ. Geol.*, 68, 612–34.

Talmage, S. B. (1925) Quantitative standards for hardness of the ore minerals, *Econ. Geol.*, 20, 535–53.

Tarkian, M. (1974) A key-diagram for the optical determination of common ore minerals, *Minerals Engng*, 6, 101–5.

Uytenbogaardt, W. and Burke, E. A. J. (1973) *Tables for Microscopic Identification of Ore Minerals.* Princeton University Press.

Vaughan, D. J. and Craig, J. R. (1978) *Mineral Chemistry of Metal Sulfides.* Cambridge University Press, England.

Vaughan, D. J. and Ixer, R. A. (1980) Studies of the sulfide mineralogy of North Pennine ores and its contribution to genetic models. Trans. Inst. Min. Metall. **89**, B99–B109.

Vokes, F. M. (1969) A review of the metamorphism of sulphide deposits, *Earth Sci. Rev.*, **5**, 99–143.

Vokes, F.M. (1973) 'Ball texture' in sulphide ores, *Geol. Foren. Stockholm Forh.*, **195**, 403–6.

Wolf, K. H. (ed.) (1976) *Handbook of Strata-bound and Stratiform Ore Deposits.* Elsevier, Amsterdam.

INDEX